BEYOND TQM

Tools & Techniques for High Performance Improvement

by

Jack L. Huffman

BEYOND TQM

Tools & Techniques for High Performance Improvement

by

Jack L. Huffman

Lanchester Press Inc
Sunnyvale CA
http://www.lanchester.com

Lanchester Press Inc.

Published in 1997
Printed in the United States of America by Patson's Press Inc., Sunnyvale, CA

Library of Congress Catalog Card Number: 97-73094

ISBN: 1-57321-012-9

Lanchester Press Inc.
P.O. Box 60621
Sunnyvale
CA 94086
http://www.lanchester.com

Layout, Lithography and Jacket Design by Costas Schuler

Dedicated to my wife and best friend, Barbara Ann, without her love and support this book would not have been possible

Contents

Section III, Teams

Section IV, Creativity Tools for Teams and Individuals

Section V, Solving Problems

Section VI, Optimizing Decisions

Section VII, Improvement Strategies and Tactics

Preface

The goal of "Unrelenting Improvement" in any organization is twofold. First, ever increasing levels of customer or client satisfaction has become a critical goal for the twenty-first century organization. Second, an ever increasing level of effectiveness for every activity in the organization continues to be a major emphasis and will escalate as natural resources become more limited. Customer satisfaction and effective use of resources drive more organizations than ever. Customers have become a critical component of decision-making and doing more with less the standard.

"Customer," as used in this book, includes any group or person affected by the organization, or groups and individuals within the organization. The definition includes employees and extends to suppliers and investors in the organization. For nonprofit and governmental organizations, the customer groups include those who supply the revenues and those who receive the benefits of the organization, including employees. "Effectiveness" means doing the right things right, with minimum use of resources provided. Effectiveness is different from "efficiency." Efficiency is a measure of maximizing outputs while minimizing inputs. To what extent the outputs are appropriate to the customers is not a factor in efficiency measures.

Another way to state the goal of this book is: The organization must continually, and relentlessly, improve all processes such that they singularly and collectively equal or exceed the Voice of Customer, in the most cost-effective manner. In short, organization must do more for their customers with fewer resources consumed.

Voice of the Customer

All customers react to how the output of a process meets their needs and expectations. The reaction may be positive, if the output meets or exceed their needs and expectations, or negative, if the output comes up short. The degree of the reaction varies considerably by person or group. Personality and mood affect the nature and degree of the reaction. The nature of the product and the time at which the customer makes a judgment, will affect the customer's reaction. Though the degree of the reaction is highly situational, what the customer requires or expects is real, and is possible to detect. Customer expectations are the "Voice of the Customer." Groups and individuals can determine the Voice of the Customer using a variety of data gathering methods outlined in Chapter One. Processes also have a voice. We "hear" the "Voice of the Process"through a variety of data-gathering methods outlined in Chapters Two through Five.

The primary purpose of this book is to help users produce a match between the Voice of the Customer and the Voice of the Process, cost-efficiently. Then, further improvement can go beyond the Voice of the Customer to produce customer delight. Do not forget that all employees are customers and have needs.

An improvement technique that combines the Voice of the Customer and the Voice of the Process will be found in Chapter Two, *Gathering and Interpreting Data: Control Charts.* One creates a picture of the process by charting the data collected from the process. This is the Voice of the Process. This "Voice" must represent a measure that is important to the customer of the process or a measure that clearly predicts the capability of the process to satisfy the customer. After the Voice of the Process is clear, the Voice of the Customer, using specifications, is added to the control chart for comparison purposes.

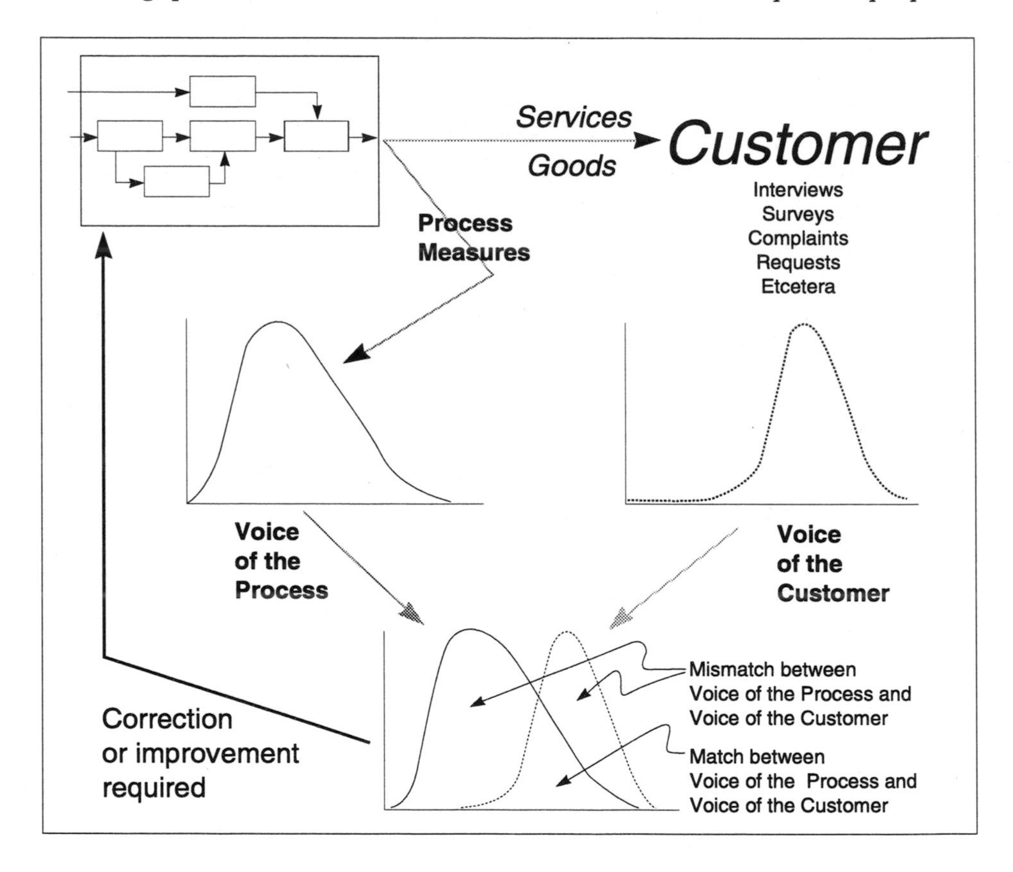

Chapter 15, *Improvement Tactics,* presents alternative approaches to the alignment of the Voice of the Process and Voice of the Customer. Precontrol (Plan-Run-Study-Act) begins by graphing the specifications of the customer. The process operator charts the same measure for the process and compares it with the customer's specifications. Improvement efforts will be directed at improving the Voice of the Process to agree with the verified Voice of the Customer. The first step is knowing that the process requires improvement. Determining how to make the necessary improvement is the next step.

This book offers an assortment of approaches, tools and tests for many improvement activities. Most, if not all, members of the organization will easily understand and learn to use the approaches and techniques outlined in this book. Nothing is reserved for management, only, or front line people, only. None of the approaches are particularly complex and some are outright simple. Simplicity assures that all employees use these improvement approaches. The simplicity of some tools and approaches should not lead one to discount their potency. Each is powerful when used properly, at the right time, in the right setting. Most are useful in a team setting and by individuals. A skilled team facilitator is highly beneficial as individuals and teams learn the appropriate application of the tools. However, team members can help each other learn their use and practice.

The Sections

The seven sections represent the seven critical initiatives organizations face as unrelenting improvement becomes the norm. Collecting data, then turning the data into information, is the focus of Section I, *Gathering and Interpreting Data.* Always making data-based decisions is a cornerstone of quality management and is consistent with the philosophy of constant improvement. Section I also includes techniques to understand processes better, and how to analyze, present and understand the data collected. This section contains most of the "seven statistical improvement tools" in common use. Chapter One contains seven methods for gathering data from processes and customers. Chapter Two provides a short course on statistical process control (SPC). Chapters Three, Four and Five present tools for better understanding processes and better understanding data.

Section II, *Directing the Improvement Effort,* aids teams and individuals in selecting the right project and then using the right improvement strategy to achieve improvement. Chapter Six offers five approaches to selecting projects that provide the potential for the greatest effect on the organization and customers of the organization. Chapter Seven offers a contemporary approach to improvement strategies: *The Four "Res."* Products, systems, processes and activities can be "Repaired" (if problems exist), "Refined" (if one wants continued improvement), "Renovated" (if there is the need for major improvement), or "Reinvented" (if the environment demands greater improvement than possible using the other three "Res").

Section III, *Teams,* includes a single chapter on team optimization. Teams of all types are one cornerstone of quality management.

Section IV, *Creativity Tools* for Teams and Individuals, also contains a single chapter. Chapter 9 deals with generating "soft" data through creative thinking, whether an individual or team.

Section V, *Solving Problems,* is appropriate for teams and individuals. A systematic problem-solving, process-improvement approach is presented in a form beneficial to any organization. Since getting to root causes is critical in solving problems, this section includes a chapter dealing with techniques for digging out root causes. Teams and individuals experience two critical problems. First, it tends to be human to jump to a conclusion before all the data concerning a problem is gathered. Then, when the problem is seemingly solved, it is also human to go on to the next problem or opportunity. And, second, most problem-solving efforts stop far short of the root cause of the problem. Section V is critical for organizations in a serious improvement initiative.

Section VI, *Optimizing Decisions,* helps both teams and individuals develop better decisions and choose among alternatives. This section also includes a chapter on testing root causes and optimum solutions. It is important that decisions are based on an identified root cause or the selection of an optimum solution will hold up to a test before full-scale implementation. The final section,

Section VII, Improvement Strategies and Tactics, includes a chapter that expands the highly valuable *plan, do, study, act* (PDSA) to make it better understood and more applicable. Chapter Fifteen applies the PDSA approach to processes not meeting customer requirements to force the process to conformance. This is a modified version of "precontrol." Chapter Fifteen also contains other improvement tactics, including the improvement of the implementation process.

A Final Thought

Other improvement and problem-solving approaches and tools exist for individuals and teams. Some are simple, but have highly limited application. Others are quite complex, and powerful, but are difficult to learn and apply.

This book provides individuals and teams the approaches and tools best suited for regular use by all members of the organization. Each approach and tool meets the test of being easily learned and applied. Most are highly effective for use by teams and individuals. Some should be part of every team meeting, whether an improvement team or not. Some should become part of every person's daily life, on the job and off. Many find use in one's personal and family life for the same reason they are useful in one's work life. These tools and approaches apply to all improvement efforts.

Always keep in mind that the content of this book is most valuable when kept simple. Do not attempt to make the tools more complex than need be. Do not use a tool for the sake of using the tool. Tools are means, not ends. The approaches help individuals and teams in choosing the right initiative, developing a focus, making progress, implementing change and maintaining gains.

Section I
Gathering and Interpreting Data

Chapter 1

Gathering and Presenting Data: Basic Approaches

Data

Optimum decisions begin with appropriate, and valid, data. However, data does not just happen. One must gather and compile data. The data must be the right data from the right sources. Generating useful data to help a team or individual in making problem-solving and process improvement will take time and energy. Once gathered, appropriate and valid data becomes the basic material for information, and it is information that leads to informed decisions.

This section offers basic approaches and techniques for gathering data efficiently and effectively. Teams or individuals may use each technique.

Measurements

Measurements are the basis for most data. Since all major decisions must be data-based, it is critical that data are collected from the right sources in the right manner. Measurements, therefore, must represent the right item or event for the measures to be useful.

Accuracy of the data is also an issue. Do not assume that the measurement is more accurate just because the measurement includes more significant figures than other measures. An oral thermometer could produce a reading of 99.4 degrees when "read" by the human eye. Using a very powerful magnifier could produce a reading of 99.43 degrees. If the thermometer is off by +.2 degree, the measurement reported to four significant figures misrepresents the accuracy of the measurement.

Averages also present a problem, if misused. Do not record the average of 4, 5 and 2 as 3.6666667, though an eight-digit calculator displays "3.6666667." The reported result of a calculation should contain no more than one additional significant figure than the number in the calculation with the least number of significant figures. In the previously mentioned example, the reported average is 3.7. Record and report the result to two significant figures since the numbers in the calculation are all accurate to one significant figure. If the numbers were 4, 5.0 and 2.00 (the 5.0 and 2.00 imply that the numbers are accurate to two and three significant figures, respectively), record the result as 3.7 since the number "4" is one significant figure. Record the average of 2.05, 5.9 and 3.117 as 3.69, not 3.689 as displayed on a calculator.

When rounding a number, if the portion of the figure being rounded off begins with the number 5 or more, round the last place remaining up to the next number. If the portion being rounded off is four or less, leave the last significant digit as is. Therefore, 36.7584 rounded off to three significant figures becomes 36.8 and 6.7546 becomes 6.75.

Several critical issues must be considered when collecting data. Measurements require resources to collect and make the necessary computations. These resources are time and money, and both are limited. The time it takes to collect data and convert the data to information delays decisions. Though erring on the side of having a little more data is best, each additional measurement comes at a cost. Collect enough data to make informed decisions, but not at the cost of a delayed decision or at a greater cost in data collection than value received. Figure 1-1 shows the relationship between how much data are collected (which depends on time) and the "quality" of the decisions that result. Quality is in quotation marks because the quality of a decision depends on many factors, including the amount of valid and timely data collected.

Figure 1-1 also shows that the accumulated quantity of data collected over time does not form a straight line. Quite a bit of informative data is collected quickly. Within a short time, enough data are collected to make a quality decision. Doubling the time allotted to data collection does not improve the quality of the decision by double. As the "amount of data collected" line suggests, the improvement in decision quality improves a small amount. Therefore, rather than allot a great deal of time collecting data, a more productive alternative is to improve the effectiveness of the data collection methods to get the second curve.

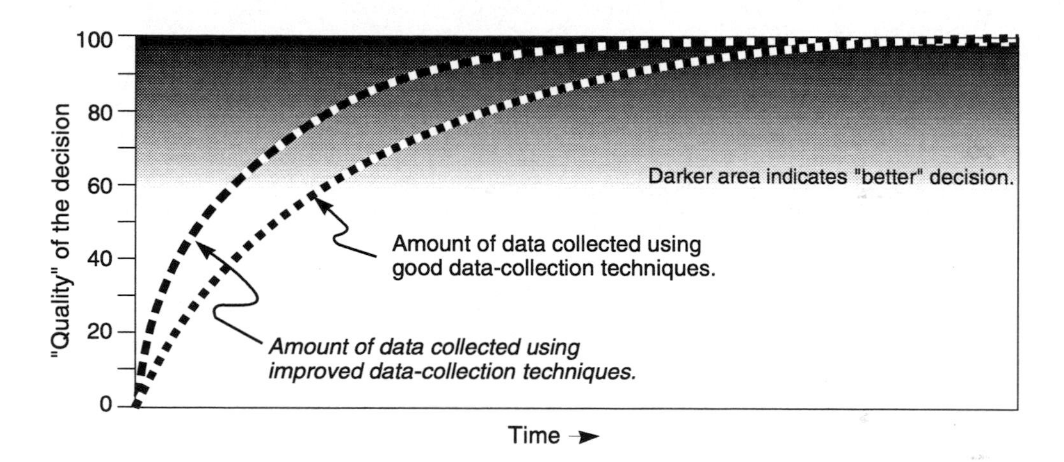

Figure 1-1 Optimizing data collection to get the maximum data in the minimum time.

Every data-collection activity should contain four considerations:

1. Data costs money and time to collect,
2. A trade off always exists between the cost to collect as much data as possible, and the "quality" of the decision based on that data,
3. An individual or team probably cannot collect all the data associated with a problem-solving or improvement initiative, and
4. Improving the data-collection methodology pays off in more and better data,earlier.

Randomness

Taking measurements at intervals or drawing samples (that is, not measuring every item or occurrence) requires the data to be collected at random. Random means that every item or occurrence has an equal chance of being chosen for measurement. Dealers always shuffle a deck of cards before dealing to players. Medicine is shaken well to make certain that the dose taken is a mixture of the contents of the bottle. A rotating barrel throughly mixes the raffle tickets to make certain each ticket has an equal chance of being drawn.

Several methods are available to produce a random sample:

1. Use a computer's random number generator. If the population of items or occurrences is large (in the hundreds or more), one could write a simple computer program to produce a set of random numbers. For example: Assume that 30 employees of 400 are to be selected at random and surveyed. Instruct the computer to provide 30 random numbers from zero to 399. Given the 30 random numbers from zero to 399, and each of 400 employees numbered from zero to 399, it is merely a matter of determining the 30 employees who match the 30 random numbers. Of course, a programmer could program the computer to do all the work and merely display the names of the employees chosen.

2. Use a random number table. Random number tables are found in most older statistics books. Since computers are much faster and easier than resorting to a random number table, random number tables are becoming obsolete. However, random number tables still have a use for quick selection of a random sample. Begin this approach by deciding how each subsequent number will be chosen after the first random number is chosen. One method is to go down the list a preselected number of positions from each previous number. Now, point to the first number on the table as randomly as possible. (Close your eyes and point at a number.) With 400 items or occurrences (employees) to draw from for the sample, consider the first or last three digits. In this example, if these three digits are 400 or more, go on to the next random number. If the number is 399 or less, it is the chosen random number. Continue until the required number of random numbers are selected.

3. Use a deck of cards. With a hundred or fewer items or occurrences in the population, use a deck of playing cards or similar cards to select samples. List each item or occurrence on each card. Shuffle the cards as in playing a game of cards. Draw the same number of cards as samples required. One approach is to write each

employee's name or employee number on a separate card. Shuffle the cards and draw 30 cards by cutting cards or another random means. Survey these 30 employees. This method works quite well in a variety of situations. The maximum number of cards is limited to the size of the deck that one can shuffle well enough to assure randomness.

4. Sample items or occurrences sequentially, according to time or number or occurrences since the last sample. In manufacturing organizations this is common in the production area. Sample every tenth unit produced, or the unit produced on the hour or at another predetermined time. Sequential sampling violates the requirement that all items have an equal chance of being selected, but the selection method is random enough that it works for process improvement. Data collectors must take care when using sequential sampling that it becomes a common occurrence for the process to produce better, or worse, output during the sampling period. A new employee, for example, might slow during the sequential sampling period to produce a higher quality output. Consider, also, the possibility that when conducting the sampling activity on the hour, something special could happen on the hour to affect the output. Sampling at one given time during the day, week or month magnifies the non-randomness of the sample.

The obvious choice for sampling is the use of a computer. Computers do not let "human" factors enter the decision such as mathematical errors, favorite times of day or week, fatigue and short cuts. The fourth method, sequential sampling, finds extensive use by organizations relying on statistical process control (SPC). Chapter 2 explains SPC.

Types of Variables

Four types of variables exist: Attribute, scaled, discrete and continuous. Each finds use in different situations. The general role-of-thumb is that continuous variables provide the best data since the data comes from standardized measuring devices or measuring conventions. Discrete variables are easier to use, but are not as accurate as continuous variables. Often, actual measurements are very difficult or impossible. Then, attribute or scaled variables become the measurement choice.

Attributes

Attribute variables take on one of two values. A statement on a "true or false" test is (or should be) either "true" or "false." No other choices exist. An attribute is an "either/or" situation. Attributes are either the easiest or most difficult data to collect. It is the easiest measurement because only two choices opposing choices exist. It is or it is not. Attributes are the most difficult data to collect because so much of what we measure in life falls on some kind of continuum. Attribute data collection assumes no gray areas. When collecting attribute data, it must be extremely clear where the response falls. In manufacturing, go/no-go gauges, standard samples and highly skilled smell, sight, sound, touch and taste senses provide decision standards. If a physical product is not involved, one must develop detailed standards to assist the data collector in determining whether the service is acceptable or not.

A critical aspect of attribute variables is that intermediate outcomes do not exist. This produces a problem during data collection. The standard must be such than any person making the decision would decide as any other person would. Most attribute measures in life are not that easily determined. A quart of milk could taste OK to one person and not OK to another. How big must a dent in a new automobile be before it is considered as a dent? Often organizations measure deliveries as on time or not on time. If a store promises delivery between 1:00 p.m. and 2:00 p.m., is 2:01 p.m. a late delivery?

Attribute data is not very "rich." Was the customer really satisfied or barely satisfied? Is the finish exquisite or marginal, but meets specifications, nevertheless? Carefully defining the attribute and developing a standard of comparison, before collecting data, is critical when using the attribute variable. People must not debate which category the measurement belongs. The categories must be completely clear, allowing no grey area.

Scaled

Scaled variables expand the attribute data into three or more categories. This results in at least three categories from which to choose, five or more being more common. Scaled variables are usually keyed to words or phrases describing the possibilities. A variation of a scaled variable that does not use words or phrases to anchor the ends of the spectrum or describe the possibilities is the "grade" approach. Educational institutions use this approach to assign grades to students. Grades typically range from "A" or "A+" to "F."

The one problem with scaled variables is the choice of words or phrases. "Good" to one person could mean "I'm completely satisfied." To another person, "Good" may mean "It barely meets my requirements." This is why so many organizations are using the "degree of agreement" approach. The survey consists of statements concerning aspects of a process output being measured. The respondent or data collector responds by choosing the phrase that best matches his or her degree of agreement or disagreement with the statement. See Exhibit 1-1 for examples of scale alternatives.

Discrete

Discrete variables make up the third form of measurement. Discrete variable measurements result in a count of incidents or increments along a scale. Time is easy to measure. Days are easy to count. People and units produced are easy to count. Customers serviced or customers complaining are easy to count. The number of reports produced are easy to count. A variety of sources record stock prices, in eights of a dollar. The key word for discrete measures is "count."

Another common factor for discrete variables is that the measures produced are a discrete set of numbers. These numbers are usually, but not limited to, whole numbers. Stock prices are an example of discrete measures that include fractions. The stock reports stock prices by eighths of a dollar. Organizations use whole dollars or tenths of a dollar to report goods or services sold. Though an item costing $23.90 and another costing $32.49 would average $28.195, organizations report the average as $28.20. Financial groups report financial data in thousands or millions of dollars.

Continuous

The fourth type of measurement is the continuous variable. This is a measurement the data collector takes and records up to the limits of the accuracy of the measurement device or convention used.

Discrete and continuous variables are often confused. Discrete measures are counts or occur at definite intervals. The number of patients admitted in a given 24-hour period is a discrete, whole number. The mean number of patients treated per day over a year is a continuous variable. It could be any number more than zero, to any number of significant figures that can be justified. People note a person's temperature in tenths of a degree. A highly accurate thermometer could record a temperature to any number of decimal places, within reason. Continuous measures could be any number, between practical limits, of course.

Individuals and teams should use discrete or continuous variables, if possible. This is because the resulting measures are generally more accurate and the data is richer. The several possible ways to report shipments are: "The last shipment was late," "I strongly disagree that the last shipment was timely," "Twenty-one shipments were late last month" or "Shipments have averaged 47 minutes late in the past 30 days." It is immediately apparent that each measure has a different degree of richness. However, each type of variable has a place in data collection. Getting information from customers, whether internal or external, usually requires attribute or scaled measures. Measures within a process provide a greater opportunity for discrete or continuous variable measures.

Examples of Variables

Attributes:

- Provide one of two alternative values to indicate process output quality
- Examples:
  - Good/No good, Go/No go, Right/Wrong
  - Satisfied/Dissatisfied
  - Delivery on time/Late delivery

Scaled Variables:

- Provide a limited range of values through comparison or opinion rather than actual measurement. For example:
  - 0 to 5, 10, or 100; or 1 to 5, or 1 to 10
  - A, B, C, D, F as in grading, but converting the results to numerical scores will be necessary. The "grade" approach functions nicely for those surveys where the respondent can relate to the assignment of grades, similar to grades given in school.
  - The use of an odd number of choices allows a center choice, if the respondent happens to be ambivalent about the question, does not care or cannot make up his or her mind. An even number of choices forces a response that is either in the positive or negative area of responses.

The service I typically receive is:

0	1	2	3	4	5
Poor	Fair	Good	Very Good	Excellent	Superb

The service I typically receive is satisfactory.

1	2	3	4	5
Strongly Disagree	Disagree	Neither Agree Nor Disagree	Agree	Strongly Agree

The grade I would assign to your overall delivery process is (circle your choice):

A+ A A- B+ B B- C+ C C- D+ D D- F

Excibit 1-1 Examples of opinion scales.

Discrete Variables:

- Provide a unit measure or predefined intervals, more directly.
- Examples:
  - Frequency of occurrence,
  - Numbers of something (a count), or
  - Dollars or dollars and cents. (Decimals and fractions are discrete if only certain ones can appear in the measures, for example: $.01 to $ .99, tenths of a gallon or stock prices.)

Continuous Variables:

- Provide a nearly infinitely variable number, or at least to accuracy of the measuring system or measuring device.
- Examples:
  - Lengths, diameters and distances,
  - Length of time,
  - Volume and weight, or
  - Efficiency, productivity and effectiveness measures.

Check Sheets

Check Sheets are useful when it is necessary to gather data as the data occur. A Check Sheet lists the possible categories (complaints, things that could go wrong, etc.) and provides space on the right side for placing an "X" or a hash mark for each occurrence. After an appropriate period checking off each category as it occurs, a history of valuable data exists for individual or team problem-solving or process improvement. Hint: Collect data until the distribution of occurrences takes on a pattern that will help in making decisions about the data set. Figures 1-2 and 1-3 show two completed Check Sheets.

Data collected on: Calls to Main Office for Repair Service From: 3/1/94 To: 3/31/94
Data collected by: John Doe

1. Late service	XX
2. Had to make second call to get it right	XXX
3. Could not finish due to lack of a part (s)	XXXXX XXXXX XXXXX X
4. Could not be fixed	XXX
5. Found no problem	XXXXX XX
6. Minor problem classified as major	XX
7. Major problem classified as minor	XXXX
8. Emergency classified as non-emergency	X
9. Non-emergency classified as emergency	XXXXX XXXXX XXXXX XXXXX XX
10. Other	XXXX

Figure 1-2 Check Sheet recording the type service request during a period of time.

Data collected on: Calls to Main Office for Repair Service From: 3/1/94 To: 3/31/94
Data collected by: John Doe

Sunday	XXXXX XX
Monday	XXXXX XXXXX XXXXX XXXXX XXXXX X
Tuesday	XXXXX XXXXX XXXXX XXX
Wednsday	XXXXX XXXXX XXX
Thursday	XXXXX XXXXX XXXXX XX
Friday	XXXXX XXXXX XXXXX XXXXX
Saturday	XXXXX XXXXX XXXX

Figure 1-3 Check Sheet recording service requests by day of week and special days.

Note that a byproduct of a Check Sheet is that it also displays the data as a frequency distribution. This is a handy feature of a Check Sheet.

Incident Charts

An Incident Chart expands the notion of a Check Sheet to show where incidents are occurring during a period, and how often. If the types of incidents occurring are of value, the chart allows expansion to show which type incidents are occurring and where they occur. The incident coding could also show other data that might help the team or expert studying the situation. Figure 1-4 is a simple Incident Chart showing where slips and falls are occurring in a building, by season. This chart directs a safety expert or accident reduction team to specific, high accident areas for further study and accident prevention. The seasons in which the slips and falls occurred also provides useful information. The focus is extremely valuable and saves quite a bit of time when determining root causes of problems and opportunities for improvement.

Using the Incident Chart shown in Figure 1-4, the team would focus their attention toward one of the two front doors, asking why three times as many slips or falls had occurred at one door. The team or expert would also consider why springtime falls had clustered in the center of the building.

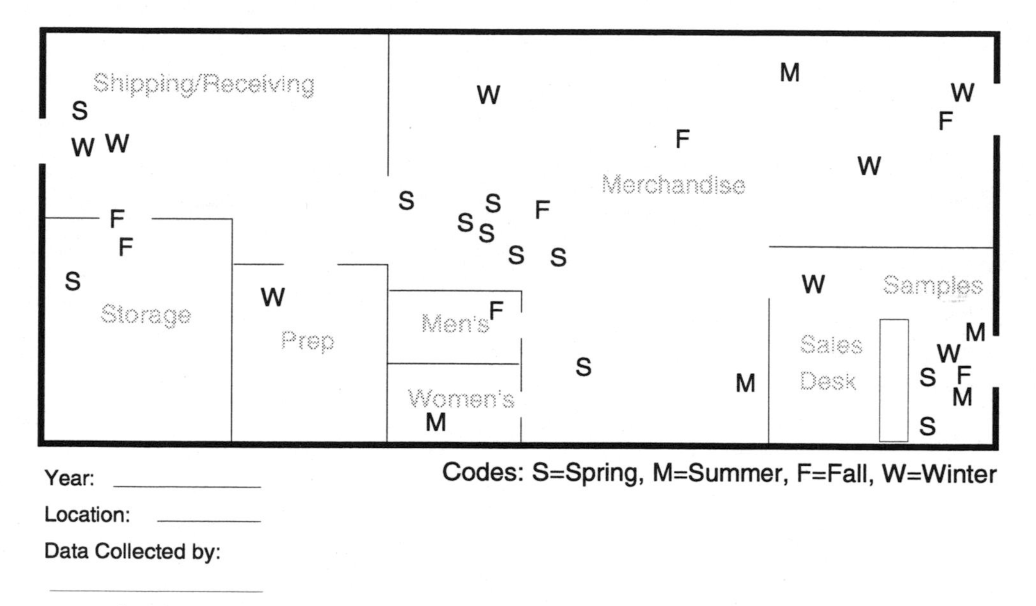

Figure 1-4 Incident Chart for locations of injury by season.

Figure 1-5 is an Incident Chart recording type of damage to cartons containing critical equipment. This chart points a packaging expert or damage reduction team toward specific types of damage to specific areas. Often, the Incident Chart, alone, identifies the cause of a problem.

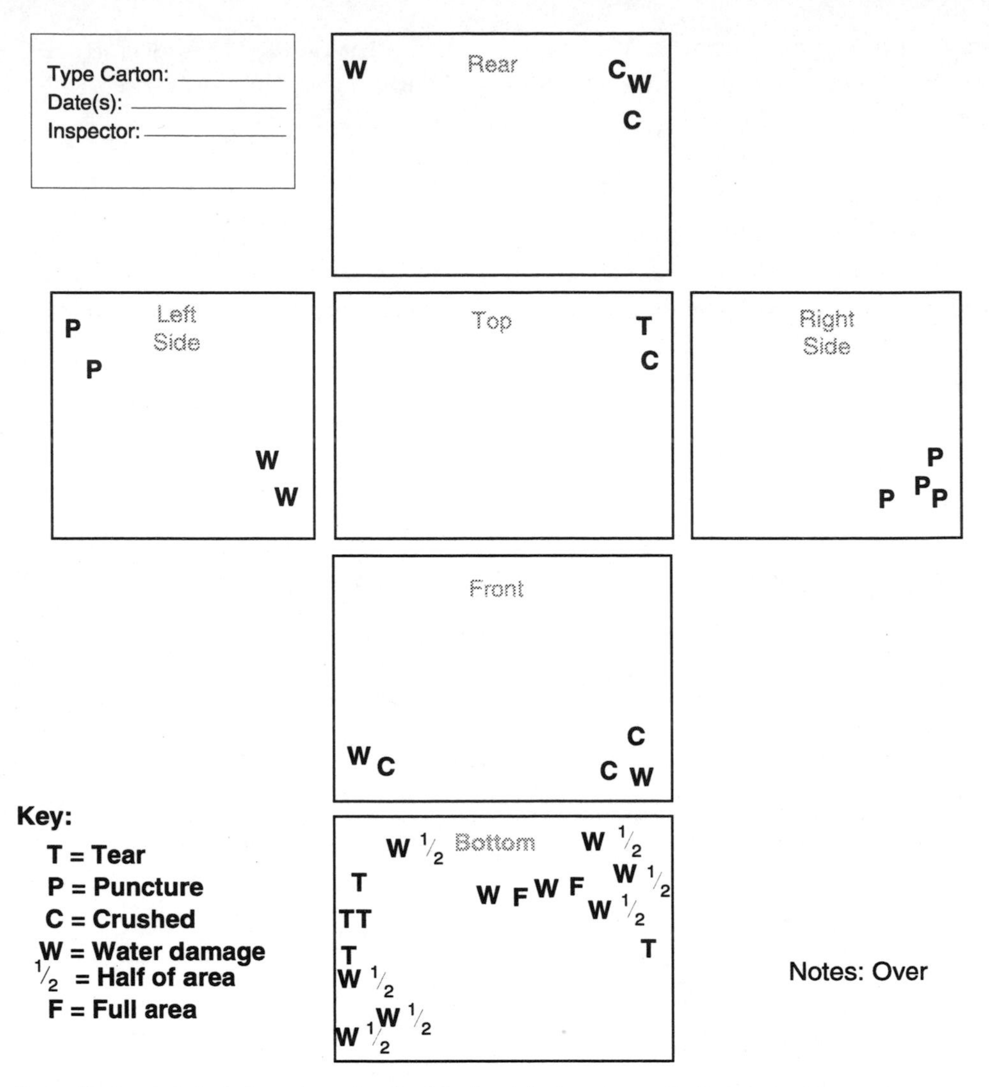

Figure 1-5 Incident Chart for carton damage.

One important thing to remember about Incident Charts is to keep the layout or diagram simple and large to allow easy data recording without data overlap. Also, there should be no question about what data to record where. If the chart requires codes or symbols, include a table of these codes or symbols on each copy of the Incident Chart.

Sun Diagram

Sun Diagrams are more useful for individuals than teams, but do find team use. The notion behind the Sun Diagram is that it provides a visual mechanism to identify all internal customers (or customer groups) of a specific person, group or process. A Sun Diagram should also show which of the customers are also suppliers.

Begin a Sun Diagram with a circle in the center of a page using the finished Sun Diagram shown below as an example. This is either a specific person, indicated by "you," or the process being studied. Add circles surrounding the center circle, one for each distinct customer type. Having twenty to forty customer types is common for an individual. Processes usually have fewer customer types, perhaps several to twenty. Processes with less than ten customer types will not benefit from a Sun Diagram.

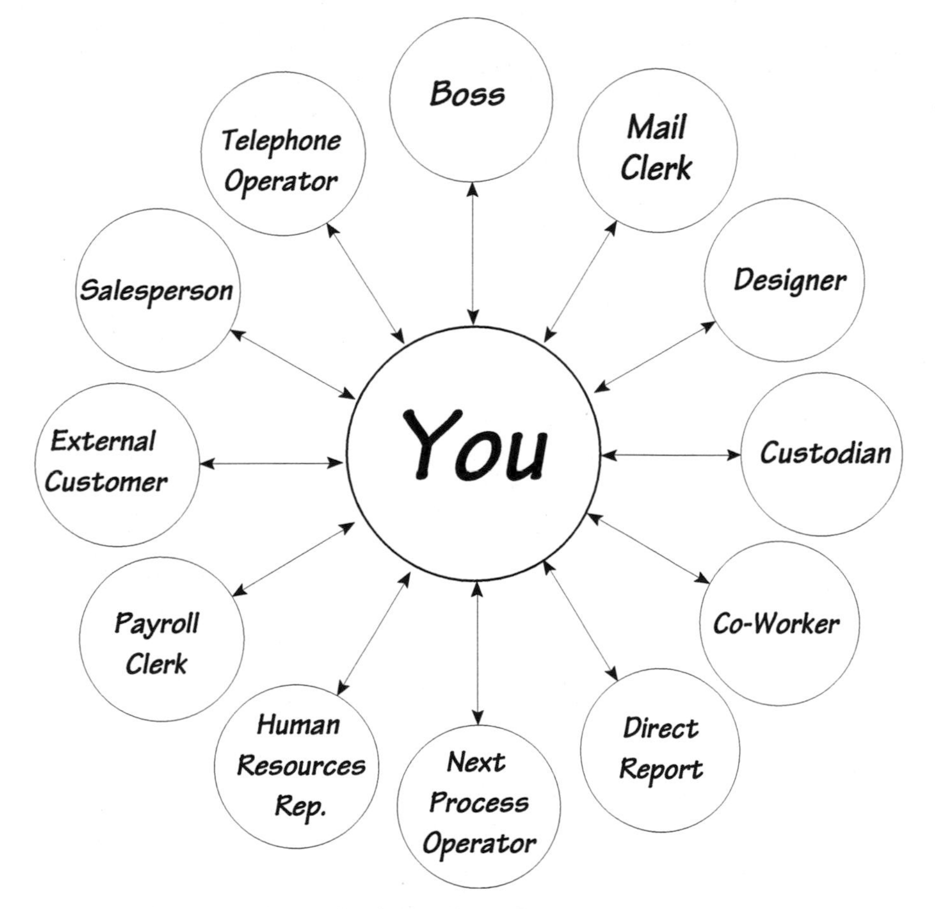

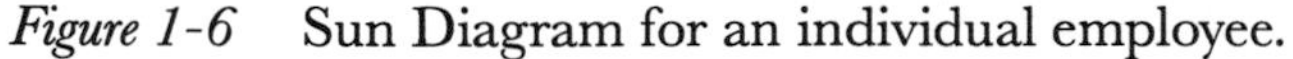

Figure 1-6 Sun Diagram for an individual employee.

After identifying all the customer types, add arrow heads to indicate the direction of goods, services or information flow. Usually, lines will have arrow heads on both ends, suggesting both a customer and supplier relationship.

When the Sun Diagram is completed, a critical customer is selected. Remember, though, that every customer is important, for a variety of reasons. Start with those causing the greatest impact or concern. At a conference with that person or group, ask three critical questions:

1. "Are any of your requirements not being met by what I (we) supply to you? What are these requirements?"
2. "Do you receive what I (we) supply to you, as you require it?"
3. "How often do you receive goods or services from me (us) that's not right the first time received? What's not right?"

These questions are critical for aligning what a person or group should supply to what internal customers require. The data collector should then add two additional questions to provide additional data for continued improvement:

1. "Is there anything that I am (we are) doing for you that is of no value to you? What?"
2. "Is there something I (we) should be doing or supplying that I am (we are) not? What?"

After these questions are asked and answers obtained, the next most critical customer is selected. This process continues until the list of critical customers is exhausted. With many customers, assume that the Pareto effect is valid. Twenty percent of the customer types should command about 80 percent of initial attention. See Chapter 4 for more information on the Pareto effect.

Sometimes the process, in its current state, can simply not meet customer requirements. Discuss this with the customer to gain understanding and buy time until process improvement is possible. Sometimes what we think is important is not really important to customers. Customers provide the best information to set priorities or improvement emphasis. Regardless, the first effort is to improve the process to meet customer requirements. Attention can then be directed to exceeding customer requirements.

Customer Questionnaires, Surveys and Interviews

Customers, whether internal or external, possess a great deal of knowledge about a problem or improvement issue. They know what went wrong (not necessarily why or how to fix it). They know what needs improvement (but, not necessarily how). When asking a customer for opinion, remember that they are giving it from their viewpoint, only. It is merely one viewpoint, not hard data. Not discounting what a customer says is important. Take it as one set of data, thank them for their time and ideas, and move on to the next customer. Move on to problem-solving or process improvement upon surveying enough customers that there is a level of comfort about the situation and the customer's perspectives.

When questioning a customer, whether in writing or by voice, be certain to ask the right questions. Make it clear that WHAT WENT WRONG, not WHO WENT WRONG is the intent of the questions. If the issue is process improvement to provide better goods or services, ask the customer to state what they typically get and what they really want.

It does not hurt to remind the customer that the price versus value relationship exists in the transaction. Sometimes meeting their expectations would cost more than they are willing to pay. Also, meeting the customer's unrealistic expectations is sometimes impossible. For example, when asked what a potential buyer would like in an automobile, the reply might be: "Heavy, fast, seats six, never breaks down and gets 50 miles per gallon." When asked how long it should take to paint the interior of a house: "Why not the same day." The only responses to these answers would be, "We would love to be able to do that, but the price would be prohibitive."

Newspaper reporters use an excellent approach to asking questions. They make certain a story is complete by making certain that what, how , when, where, who and why have been answered. The same approach can be used to gather information from customers.

- "WHAT went wrong or keeps going wrong?"
- "HOW does it seem to happen?"
- "WHEN did or does it happen?"
- "WHERE did or does it happen?"
- "WHO was involved?" (Again, not necessarily to identify who is creating the problem, but to assist in identifying problem processes or to pinpoint the cause of the problem.)
- "Do you have any ideas WHY it happens?"

Interviewers need details to explore the causes of problems and therefore prevent the problem from occurring again. One last point: Part of everybody's job as a problem solver is to separate fact from fiction and realistic expectations from unrealistic expectations.

Surveys

Surveys range from simple, one-on-one interviews to extensive questionnaires involving thousands of people, requiring experts to conduct and analyze. For problem-solving and process improvement teams, sticking to a simple approach is best. Four to six targeted questions can gather a great deal of information from most of the customers asked, whether internal or external. A few open questions, those requesting comments or observations from the respondent, will provide extremely fertile information. Often surveys provide large amounts of subjective information that does not seem to provide direction. Chapter 5 presents "Clustering," an approach to analysing large amounts of subjective data.

Several critical issues should be considered when constructing and conducting a written survey:

1. Fewer questions generally produce more returned surveys,
2. More questions provide greater and richer data,
3. Using the wrong words for possible responses (exactly what does "good" or "fast" mean?) will contaminate the data,
4. Leading the respondent to responding in a certain way by carefully wording the question in that direction is easy ("When did we begin disappointing you?"),
5. Users of the survey data must trust responses from a well-designed survey, properly conducted and analyzed,
6. Use a random sample for large customer bases,
7. Focus the questions, and
8. Surveys provide a baseline against which a later survey (preferably the same survey) will suggest how much change can be attributed to improvement efforts.

Simple surveys are easy to design if one follows certain rules-of-thumb. Consider these concerns when designing a survey:

1. **Determine what you need to know**. Do not ask that which you really do not need to know or will not use, or where the answer is obvious.

   EXAMPLES OF USELESS QUESTIONS:

   - "Is a lower price at the same quality important to you?"
   - "Do you enjoy being mistreated?"

2. **Ask for quantitative data**, if possible.

   EXAMPLES:

   - "How many days did it take to deliver your last order?" ________
   - "How many times in the last week did we not deliver to your expectations?"_______

3. **Specific response ranges.**

   EXAMPLE:

   - "How long did it take us to respond to your most recent request?"

     ❑ Less than two hours ❑ Same day ❑ Next day ❑ Two or more days

4. **Respondents can be asked to select a critical few from a large list of possibilities.**

   EXAMPLE:

   - Circle the four most important characteristics of an office for you:

Larger size than current office	Has a door that locks	Has an outside window
Allows for a variety of arrangements	Has inside window	Allows for small meetings
Square	Rectangle	Long rectangle
Odd shape	Modular design	Shared: space for two

Other ideas: ______________________________

____

____

5. **The order of the questions or alternatives is important and will affect the results.**

   CONSIDER THESE EFFECTS:

   a. Respondents will not give each question equal consideration and a previous question will affect the next. A larger number of questions magnify this effect on the survey.
   b. Respondents will often select different alternatives based on the order of the alternatives.

   Therefore:

   c. Keep the number of questions to a minimum, four or five to maximize return and reduce the order effect.
   d. Test several forms of the survey to learn the extent of the order effect.
   e. To compare future survey data with current data, retain the same questions in the same order with the same ordered response alternatives. However, equate this with the advantages of continued improvement of the survey. Changes in question wording and question alternatives may improve the subsequent data but eliminate comparison to the baseline data.

FOR MORE ON SURVEYS:

Fink, A. & Kosecoff, J. How to Conduct Surveys: A Step by Step Guide. Newbury, Calif.: Sage Publications, 1985.

Alreck, P. L. & Settle, R. B. The Survey Research Handbook , Second Edition, Chicago: Irwin Professional Publishing, 1995.

Sudman, S. & Bradburn, N. Asking Questions: A Practical Guide to Questionnaire Design. San Francisco, Calif: Jossey-Bass, 1982.

Focus Groups

Customer opinion mechanisms based on direct feedback, interviews and surveys have one shortcoming: The feedback directly relates to what the supplier expects. These forms of customer feedback do not explore what the customers may really want and need. Questionnaires may not explore what customers really need, but do not know they need. The Focus Group approach permits customers and potential customers the opportunity to explore the situation, in-depth, and “freewheel.”

Marketing groups often use the Focus Group approach to gather softer, but richer, data about products and potential products. Typically, the Focus Group moderator will bring together a group of eight to twelve individuals for an hour or two. A moderator will lead the group through a discussion of selected topics. The moderator guides the focus group to areas of concern and allows the individual to discuss and scrutinize the topic at their leisure. As in brainstorming, one member will often build on another member's comment, or may take a surprising, and yet worthwhile tangent. The moderator collects notes, maintains some appearance of order, and moves the group to the next topic at the appropriate time. Moderators of marketing focus groups are good sources of improvement Focus Group moderators. Marketing focus groups are audio-recorded or video-taped to allow further analysis of the discussion. Recording may be appropriate for improvement Focus Groups, if the moderator can maintain group openness. Problem-solving and improvement efforts can benefit from the Focus Group approach. Whereas the marketing focus group members are generally strangers, internal Focus Group members may be well-acquainted. If the group gets into emotional issues, individuals may not be open with each other. The problem of group-think may emerge if a higher ranking or charismatic individual expresses an opinion. The moderator must be aware of these possible problems and challenge the group to express their own opinions.

A Focus Group session need not be as professionally conducted as does a marketing focus group session, but some basic steps are useful.

Planning:

1. SELECT THE GROUP: Select the focus group participants as an improvement team would be selected. Create a diverse group of six to twelve people.
2. SELECT A MODERATOR: The moderator could be a focus group moderator from marketing, a quality improvement facilitator, or another person perceived as dissociated from the topic. An additional person acting as scribe is highly beneficial.
3. SCHEDULE FOCUS GROUP MEETING: Bring the group together in isolation for an hour and a half to two hours. Set up the room as informal as possible. Seat everyone around one large table or seat the individuals informally in chairs, without a table. The room should contain a flip chart and a method to display completed pages from the flip chart. Refreshments help.

The focus group meeting:

4. INTRODUCTION: The moderator thanks the group for their time, briefly notes the ground rules (see Exhibits 1-2 and 1-3) and mentions the topic or issue.
5. OPENING QUESTION OR CHALLENGE: "What is right?" The moderator could begin the session by asking the group for their opinion about what is right, what works, what satisfies them. For example, if the focus group is addressing office layouts for an office layout improvement team, the moderator could ask, "What works right about our current office layout?" Another form of this question could be "What do you like about our current office arrangement?" Or, "What must we retain?" The moderator permits the group to talk about what is right long enough to collect sufficient data on what could or should be retained. Disagreements among the members may occur. That is normal and provides fertile information.
6. WHAT IMPROVEMENTS ARE NEEDED? The moderator then asks about that which needs improvement. Are there suggestions? The moderator needs to be particularly vigilant during this part of the session. The group can easily get far from the subject or go far beyond the practical restraints of budget, time or practicality. At one point the moderator must decide to move on to the next focus of the discussion.
7. WHAT ARE WE MISSING? Ask the group to comment on what might be done or included which is currently missing. Often, private agendas begin surfacing here. Extroverts with a strong need may overpower other members of the focus group. The moderator must remain in control without affecting the open thinking of the group.
8. THANK THE GROUP MEMBERS for their time, interest and ideas.

Post meeting:

9. MODERATOR AND SCRIBE CAPTURE NUANCES FROM THE MEETING. Often the nuances or undercurrents from the meeting are worth noting. Capture these with the realization that they are not data or information from the focus group. However, this information could be valuable to the improvement team as they gather additional information or try to understand a situation better.

Internal Focus Group Rules

1. We are here to gather information about a problem or improvement opportunity, not to solve the problem or recommend an improvement.
2. We will record the information discussed on a flip chart so that everybody may see what we will take from the room. This session will not be recorded.
3. Everyone will be given the opportunity to talk.
4. All comments and suggestions are welcome.
5. All comments made will be anonymous. Who said what will not be noted.
6. We should keep arguments to an absolute minimum. We are here to collect information from you and to get your ideas, not to air or settle disagreements.

Exhibit 1-2 Rules for an internal Focus Group.

External Focus Group Rules

1. This is not a marketing focus group. We are here to gather information about a problem or improvement opportunity.
2. The information discussed will be recorded on a flip chart so that everybody may see what we will take from the room. This session will not be audio taped or videotaped.
3. Everyone will be given the opportunity to talk.
4. All comments and suggestions are welcome.
5. All comments made will be anonymous. We will not note who said what.
6. We are here to collect information from you and to get your ideas, not to solve problems directly.

Exhibit 1-3 Rules for an external Focus Group.

Chapter 2

Gathering, Interpreting and Presenting Data: Control Charts

Control Charts

Control Charts make up a special category of data collection and presentation methods. These charts not only provide a mechanism to record the data, but also apply simplified statistical approaches to present the data in a way that assists decision-making efforts. They present a pictorial understanding of a key measure of a process. Control Charts show the "health" of a process in the key measure being charted, and permit continuous improvement and maintenance of those improvements.

Control Charts

Control Charts are a variety of charts used in the application of statistical process control (SPC). This section will focus on two categories of charts found most useful in any type organization: The first is the Variables Chart, which displays averages of measurable values and the range of the measures for a sample. The second is the Attributes Chart, which displays number of defects or defectives, or the proportion of defects or defectives. Other Control Charts with more limited use can be found in statistical process control guides.

Using a Control Chart is one of the most useful approaches to gaining an understanding of a process. The Control Chart then allows the operator to confirm that improvement has taken place and make certain that the process remains improved. Finally, the Control Chart alerts the operator to subsequent undesired changes in the process. Essentially, a Control Chart visually displays a critical measure of a process in a form that helps the process operator in understanding how the process is behaving, but only for the charted measure. For example, if a physical characteristic or amount of time consumed is critical to the process customer, the process operator can measure at appropriate intervals and chart that characteristic. After plotting twenty or more measures, the process operator calculates the upper and lower control limits and adds these to the Control Chart. These control limits provide the basis for making decisions about process behavior. If the process suggests an out-of-control condition, problem solving activities can take place to bring the process into control. Once in control, the process operator begins making or suggesting improvements to exceed customer needs. Standardization and maintenance of changes that improve processes should follow improvements. Corrective actions should follow changes that do not improve, or even degrade process outputs.

The differentiation between special and common causes is critical to Control Charts. See Chapter 15, *Improvement Tactics*, for an in-depth discussion of actions on systems for common causes of variation and local actions to eliminate special causes of problems. Stated briefly, an individual or team can track down something specifically causing the variation of process output. Common causes of variation are natural causes of variation and are reduced only by making general changes to the process. Common cause variation occurs strictly at random. A process void of special causes of variation, is considered stable and will produce data points distributed, over time, between the control limits to form a normal distribution. A critical issue is that in-control processes are predictable, whether or not they meet the needs of the process customer. Plotting the data from an out-of-control process results in a graph exhibiting strange patterns of variation. Points often fall outside the control lines. Sequences of points may fall above or below the average, only. The points may rise or fall over several periods. In short, the pattern is not random and does not form a normal distribution (see Chapter 4). The last portion of this section offers several decision rules for determining that a process is out of control or has gone out of control after having been in control.

Remember, if a process is out of control, it is unpredictable. Figure 2-1 illustrates a process in control, suggesting the effects of common variation, only. Figure 2-2 illustrates a process out of control. Common causes of variation are masked by the special causes, shown by shifts, sudden large variations and trends.

Variation from an out-of-control situation may mask any actions aimed at producing process improvement. Worse yet, an inappropriate change may be more than offset by special causes acting in that moment in the opposite direction. A high risk of standardizing a hindrance to a process or discarding an improvement not shown to have improved the process exists when improving an out-of-control process, rather than correcting special problems. Experts strongly emphasize that problem-solving efforts brings the process under control, then improvement efforts will improve the process. If the process shown in Figure 2-1 subsequently goes out of control, or the average moves up or down, it will be readily apparent. If a special cause creeps in, it will also be readily apparent from the abnormal fluctuations.

Note that the variation in Figure 2-2 seems unusual. The defect rate is high for a time before becoming lower. Then the defect rate bounces between high and low, and finally increases again after being low for a few periods. In a manufacturing situation, no certain way exists to know how many units the process needs to produce at any given future date to make up for defective units. In a service situation, the customer service people, for example, would never know what the next day will produce. In both situations, one would have no way of knowing what the processes would produce the next day.

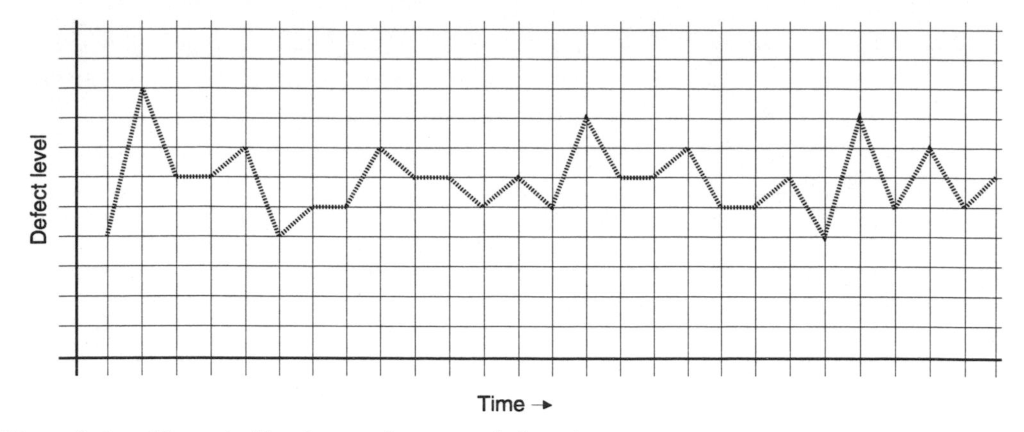

Figure 2-1 Chart indicating an in-control situation.

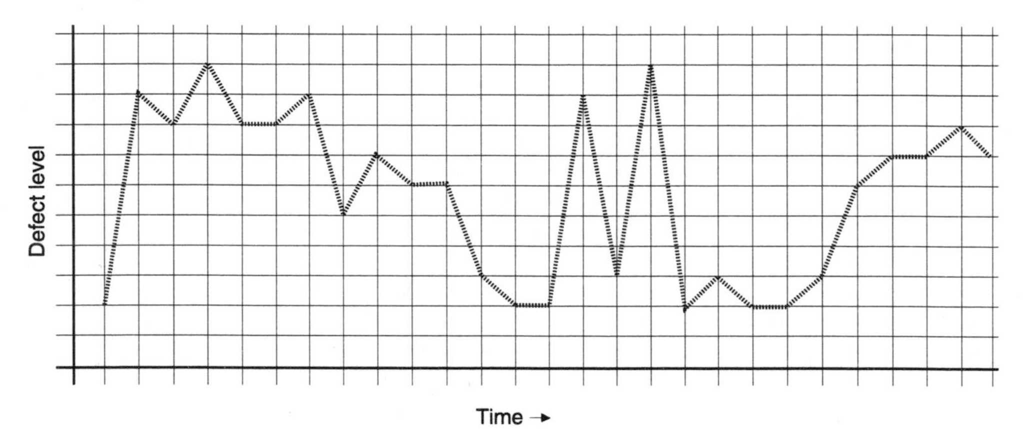

Figure 2-2 Chart indicating an out-of-control situation.

When a process is in control, variation still exists, but the variation is predictable. Much of the variation will be within a small range near the average for the process.

Individuals and teams apply Control Charts to processes in four phases. During the first phase, *the capability study*, the process operator runs the process *without* any adjustments or changes to establish the degree of process output predictability and process capability. The second phase, *problem-solving feedback*, provides feedback as the operator removes special causes. After the removal of all special causes and the process is under control, the operator begins the third phase, *improvement feedback*. The operator, an expert or a team develops improvements to make the process better meet or exceed customer requirements. Another aspect of this phase of process improvement is to reduce scrap or improve process efficiency. This critical third phase of Control Charting provides accurate responses concerning the degree of process improvement. The fourth phase of

Control Charting, *process maintenance,* not only provides confirmation that a process continues to run as expected, but also provides early warnings of problems. These warnings may suggest that the process could be going out of control, that special causes are creeping in, or that the mean is shifting. Exhibit 2-1 provides a summary of the four phases.

Control Chart Phases

Phase One: Capability Study

Use the Control Chart to learn the current capability of the process. During this phase the process operator must allow the process to run as it normally does. The process operator or management should not make any changes to the process or inputs to the process, and must not tamper with the process in an attempt to make the output measures appear better than usual. After 20 to 30 measurement periods, the chart will suggest the capability of the process, unless variation is due to cycles longer than the time that the 20 to 30 periods cover.

Phase Two: Problem-Solving Feedback

The second phase provides information on the amount of improvement because of specific problem elimination initiatives. Carrying out one solution at a time is critical to being able to trace the Control Chart changes back to specific improvements. This phase should continue until the process is in control and, therefore, will produce a predictable output for the measure charted.

Phase Three: Improvement Feedback

As with the application of corrections, make only one improvement at a time to allow changes in the Control Chart to be attributable to a specific improvement. Continue making process improvements until the charted measure consistently meets or exceeds customer expectations.

Phase Four: Process Maintenance

Continued use of the Control Chart provides feedback about the "health" of the process. During this phase the sampling period can be extended unless:

1. The charted measure is a critical measure of a critical process, or
2. Any change in the process could dramatically affect the customer.

Exhibit 2-1 The four phases of Control Charts.

Control Chart types

Two types of Control Charts are useful in most organizations: the Variables Chart and the Attributes Chart. Use a Variables Chart when the measurement is a continuous or discrete variable, as with dimensional and time measurements. In some circumstances, using a scaled variable to produce a Variables Control Chart is possible. The problem with the scaled variable is the range of values is usually limited to five to ten discrete values. Though better than nothing, a Variables Control Chart using scaled variables lacks the discrimination of process change. A run chart plotting sample averages and standard deviation over time is another choice for scaled variables.

Discrete and continuous variables provide a richness of data and should be the first choice for process measurement. Often, though, taking a variable measurement is difficult or impossible. Perhaps it is only possible to determine that the critical output aspect is either "right" or "wrong." Nothing exists between right and wrong. The output meets or does not meet customer requirements. It is "good" or "not good." Manufacturing groups use the terms "go" and "no go." In a physical inspection, a defect or defects exist or do not exist. For these "yes/no" situations, an Attributes Chart is the choice.

Variables Charts

Two types of Variables Charts are of value to many organizations: XBar&R and Individual Samples (X&MR). An XBar&R chart is appropriate when taking two or more samples is possible. The application of XBar&R charts is widespread in manufacturing organizations where larger quantities of output exist for ongoing processes. An X&MR Chart is similar to the XBar&R except that the operator takes only one sample. Both have wide applications in goods and services industries, and institutions.

The XBar&R chart is gaining popularity in service organizations as they discover measures for process output. The chart gets its name from the fact that it displays the averages (X-bar or $\overline{X}$) of groups of measurements, along with the ranges (R) of those measurements. Most organizations find that five sample measurements makes the best compromises between enough measurements to get an average that represents the period or lot, and yet is not too large as to consume excessive amounts of time making the measurements.

The process operator makes measurements for 20 to 30 periods intervals and records these measurements on the XBar&R Chart. Twenty to thirty periods or lots will provide a reasonable representation of how the process is operating. At this point the chart appears similar to a run chart except that two graphs are evident, one for the averages and another for the ranges. Compute the average of the averages ($\overline{\overline{X}}$ or "grand average"), upper and lower control limits of the averages ($UCL_{\overline{X}}$ and $LCL_{\overline{X}}$), the average of the ranges ($\overline{R}$), and the upper and lower control limits of the ranges (UCL_R and LCL_R). Draw the lines representing these values on the chart. This completes the XBar&R Chart for the first phase of process improvement activity.

Variables Control Charts differ from run charts for three primary reasons. First, each point represents more than a single measurement. Each point in the averages graph represents the average of several measurements from a period or a lot. Each point in the ranges graph represents the difference between the smallest value and the largest. Second, ranges present the variability within the individual sample lots, or a single measurement compared with the previous measurement. Third, control limits aid in determining process stability and capability.

The second type of Variables Control Chart is the Individual Samples Chart or X&MR Chart. Often, processes do not produce the quantity of output that would permit the operator to take more than one sample measurement at a time. If a process produces only one output per day, taking more than one sample per day would be impossible. The Individual Samples Chart is also useful if making the measurement destroys the sample, as in drop-testing a notebook computer to test the continuing ability to survive a specified drop, as designed. Unfortunately, with only one sample, computing the range of measurements between the samples is impossible. Since the control limits are computed using a formula that depends on an average range value, the Individual Samples Chart plots a range value by computing the difference between the latest sample measurement and the previous measurement. The resulting X&MR Chart is not as "rich" in information as a XBar&R Chart, but the chart will provide feedback for decisions about process stability, improvement and continued control.

Attributes Charts

Often, making a discrete measurement is not possible. Assigning a scaled value to the output measurement may even be impossible. In these cases, making an attribute measurement is the last resort. One does not really "measure" attributes. Standards suggest attribute measures. The aspect being measured conforms to the standard or it does not. The attribute values are only useful in an Attributes Chart or a Run Chart.

Four different types of Attribute Charts are available. Which to use depends on whether the study tracks single defects or defective output units, and whether or not the sample size is constant. The "p" chart is useful if the sample size varies and each defect is important to retain in the data. The p chart uses the **p**ercentage of defects found per sample. The "np" chart is appropriate for constant sample sizes where each sample is defective or not. The "np" chart graphs the **n**umber of samples (**p**arts) with one or more defects. Since the chart does not retain the fact that two or more defects could occur per sample, multiple defect information is lost. The "c" chart graphs the individual nonconformances (the defect **c**ounts). The "u" chart graphs the average number of defects per sample (**u**nit) in the group sampled. The u chart is useful for constant sample sizes, only. Exhibit 2-2 outlines the four different types and the application of each.

Additional Information

For more information about the application of Statistical Process Control refer to:

Amsden, R. T., Butler, H. E. & Amsden, D. M. *SPC Simplified, Practical Steps to Quality.* White Plains, NY: Quality Resources, 1989.

Amsden, D. M., Butler, H. E. & Amsden, R. T. *SPC Simplified for Services, Practical Tools for Continuous Improvement.* White Plains, NY: Quality Resources, 1991.

Fundamental Statistical Process Control Reference Manual. Southfield, MI: Automotive Industry Action Group (AIAG) 1991. This manual is available from the AIAG by calling (313) 358-3570.

p Charts:

- Graphs percentage (or proportion) of samples not conforming
- Remember that the "p" stands for "percentage" or "proportion" of samples found not conforming in the group of samples taken
- Each sample period or lot need not be of the same size (applicable to varying lot sizes or process output rates)
- Typically used when only one defect can occur per sample and lot sizes can vary
- Use the p Chart if more than one defect can occur per sample, but the defects are collectively considered no worse than a single defect
- Example: A clerk samples five handwritten orders. Three lack good legibility. The clerk records "60" because three sample orders of five (60%) were defective. In the next period, the sales people took many more orders and the clerk sampled ten handwritten orders. Four lacked good legibility. The clerk records "40" for the next period (four defective orders out of ten or 40%).

np Charts:

- Graphs the number of samples in the sample lot or period not conforming
- Remember that the "np" stands for "number of parts or pieces" not conforming
- Sample lot size must be constant
- Used when only one defect can occur per sample, the lot sizes are constant and the actual number of sample cases not conforming is valuable information
- Use the np Chart if more than one defect can occur per sample, but the defects are collectively considered no worse than a single defect
- Example: A clerk samples five handwritten orders. One order contains three errors and another contains four errors. The other three were OK. The clerk records "2" because two of the sample orders were defective. The next period or lot of orders must have five orders sampled. The clerk records number of samples defective even if some samples contain more than one defect.

c Chart:

- The c Chart displays the number of individual nonconformances in the sample period or sample lot
- Remember that "c" stands for "total defect counts"
- Sample lot size must be constant
- Used when more than one defect can occur in each sample and each defect is important
- Example: A clerk samples five handwritten orders. One contains three errors and another containes four errors. The other three were OK. The clerk records "7" because seven different defects were found. The clerk must continue to sample five orders during subsequent periods or from future lots and record the number of errors.

u Chart:

- The u Chart displays the average number of nonconformities per sample
- Remember that "u" stands for "defects per unit" and a unit is an individual sample case
- Sample lot size need not be constant (allows varying lot sizes or process output rates)
- Used when more than one defect can occur in each sample and keeping track of the number of defects in each sample is important for improvement purposes.
- Example: A clerk samples five handwritten letters. One contains three errors and another contains five errors. The other three are OK. The clerk records "1.6" because eight errors were found in five samples (eight divided by five equals 1.6). One could sample any reasonable number of orders during the next period, depending on the quantity produced during the period or the lot size.

Exhibit 2-2 Types of Attribute Charts and their uses.

Control Charting a Process

Teams and individuals bring processes in control, improve and then maintain the processes, following the same steps, whether using a Variables, Individuals or Attributes Chart. What is to measure is chosen in Step One. The chosen measure must suggest the "health" of the process in forms of the customer's expectations. The good or service being measured provides the basis for deciding how to measure it, the quantity to measure and how often to make the measurement. This naturally leads to the type of Control Chart required.

Having decided what to measure, the process is ready for measurements. As previously mentioned, it is extremely critical that the process operator allow the process to run without intervention or tampering. Any external changes to the process such as changing the process operators, adjustments to make the measure look better, closer management attention, or whatever, will contaminate the results. Contaminated results will mask the effects of corrective actions and improvement.

Record the data on the chart, directly. An alternative at this point is to automate the data collection activity. Often, automated systems already contain the required data. For example, this is frequently true of financial data, sales order input and delivery data, time related data, and a variety of other data contained in an organization's computerized record system. It is also appropriate to mention at this point that an assortment of software packages are available to automate the actual charting. Employees should learn to chart a process manually to gain insight on how the data are collected and charted. This develops an appreciation for the data and the chart.

The first phase of process charting provides insight to process capability. Repeatedly within organizations, many assume that process outputs are delivering much more than the process can actually deliver. Sometimes promising a better output than possible from the process is an attempt to "satisfy" the customer before delivery. An excellent example is the delivery of goods such as furniture. A furniture store promises to the ultimate customer, the consumer, that delivery will occur on a given Thursday between noon and one o'clock. Most delivery processes in a larger city cannot deliver within a one-hour window. Too many internal and external influences affect delivery. The product availability affects delivery. Then the goods must be loaded along with other goods for delivery. The driver must cope with the variables of traffic and unloading other customer's purchases. Finally, the unknown situation of each customer's home or facility adds variability to the delivery process. Though the specification given to the customer is

12:30 p.m. plus or minus thirty minutes, the process may only be capable of delivery at 12:30 p.m. plus or minus two hours. Understanding current process capability, with real data, is critical to process correction and improvement, and the resulting improvement in customer satisfaction.

Control Charting provides the necessary information, in pictorial form, to understanding how the process is functioning and the results of corrections and improvements, backed by data. A Control Chart is one method useful in converting data into information.

Process charting

The Control Charting process consists of six steps that the person charting the process must be follow in sequence. The person charting the process must be very careful not to skip Step Four, "Collect the data without process change or tampering." Any changes purposely made to the process during Step Four will mask the variation in the process, and will probably increase the variation. The other critical step is Step Six, "Determine process stage and act accordingly." Improvement takes place during this step.

Step One: Choose what is to be measured:

- This must be a critical measure of process health important to the customer, whether the next process or the ultimate consumer.

Step Two: Choose the measurement interval and the number of samples in each interval:

- Refer to "Measurements," Chapter One.

Step Three: Choose the type chart:

- Variable with two or more samples per period or lot: XBar&R Chart
- Variable with only one sample per period or lot: Individuals Chart
- Attribute, constant sample/lot size and each sample is defective or not despite the number of defects in any single sample: np Chart
- Attribute, constant sample/lot size and individual samples may have more than one defect: c Chart
- Attribute, varying sample/lot size and each sample is defective or not: p Chart
- Attribute, varying sample/lot size and each sample may have more than one defect and each defect is important to consider: u Chart

Step Four: Collect the data without process change or tampering:

- Make the measurements over a period of 20-30 periods or lots
- Recording the data directly to the chart is permissible
- Permit the process to operate as it would normally operate. No attempt to force the process to perform better than normal should take place as this would mask the actual capability of the process.

Step Five: Complete the chart:

- See the appropriate chart type later in this chapter

Step Six: Determine process stage and act accordingly:

1st Stage: Out of control, not predictable?

- Address special causes to bring the process into control
- Compute new limits and grand averages
- Continue to chart.

2nd Stage: Process average (average of the averages) beyond acceptable limits?

- Adjust the process and continue to chart

3rd Stage: Process in control but not meeting requirements of the customer?

- Make improvements and keep on charting.

4th Stage: Process in control but further improvements will not improve the process to any degree?

- If further improvement is critical for customer satisfaction, resort to a Renovation or Reinvention strategy (see Chapter Seven). Otherwise, keep on charting.

5th Stage: Process in control but needs efficiency improvement or made to be more robust?

- Make improvements and continue charting to confirm the process continues to meet customer needs.
- Reduce or eliminate those factors that produce common causes of variation.
- Make improvements that bring the control limits well within the customers' requirements of the process.

6th Stage: Process in control, meeting, often exceeding customer requirements.

- Continue to chart the measure (relaxing the sampling frequency) to confirm that the process has not changed and to provide early warning of process problems.

Control Charts and specifications

As unusual as it may seem, a direct relationship between control limits and specification limits does not exist. Formulas determine control limits, based on sample lot measurements. Specification limits are set by experts, standards organizations, industry practices or by customers of the process. The same holds true for the process grand average and the specified target value.

After the control limits and specification limits are added to the chart, the person completing the chart can determine the degree to which the process will meet or exceed process requirements. If either specification limit is within the control limits, the process will not consistently meet process requirements. If both specification limits are within the process specifications, the situation is even worse.

When the specification limits fall outside the control limits for a process in control, the process will produce conforming output on a highly consistent basis. This is true only as long as the process remains in control. To learn when the process is going or has gone out of control, see the later portion of the chapter, "Actions on special causes."

The amount that the specification limits should be beyond the control limits is one-sixth the distance between the lower and upper control limits. Figure 2-3 represents a process in control, with specification limits the one-sixth distance beyond the control limits. This assures conforming sample values most of the time. In the example, the specification limits are at the plus and minus four sigma limits. Since the formulas for control limits

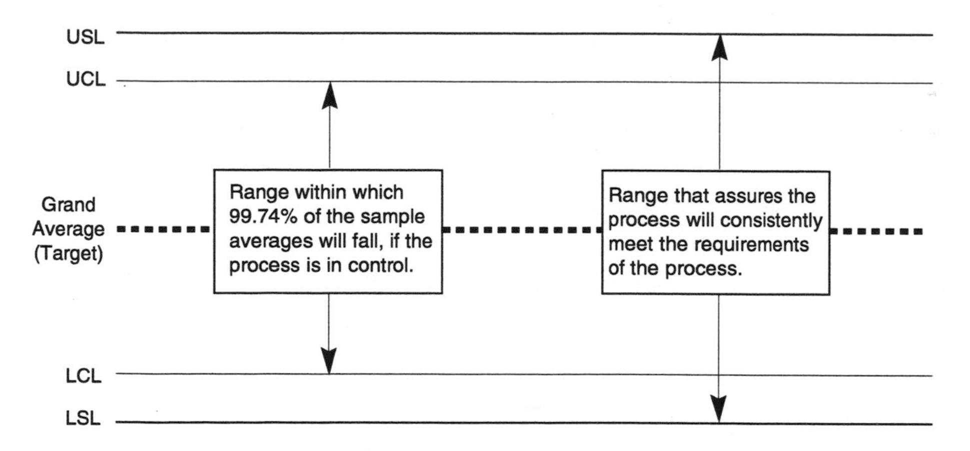

Figure 2-3 Control and specification limits, grand average and target values for a process in control and capable of producing an output which consistently falls with the specification limits.

produce values representing three standard deviations (sigma) above and below the grand average, specification limits at the control limits produce a "three sigma" process. In Figure 2-5, the specification limits are one additional sigma beyond the control limits. The example process is a "four sigma" process. Some critical processes must be well within the specification limits. For example, landing a commercial aircraft should be a "six sigma" process. Six sigma processes produce outputs with the sampled measure falling outside the specification limits only about three times in a million samples. The operator can extrapolate this to mean that the process will produce a conforming output with only three defects per million.

The relationship that exists between control limits and specification limits also exists between the process grand average and the specified target value. Differences between the grand average and target value indicate to what extent the process is off-center. A process in control and within control limits and specifications limits having the same spread, could produce considerable levels of defects if the grand average has shifted excessively from the specified target. A process will produce highly conforming output measurements only if there is alignment between the grand average and target value, and the specification limits are beyond the control limits.

Standard Deviation or Sigma

Readers need not understand the concept of standard deviation or sigma to understand the material in this handbook. However, an appreciation for the notion behind standard deviation is helpful in the struggle to understand the concept as it applies to Control Charts. Also, the terms "sigma" and "standard deviation" are becoming more common within organizations using statistical approaches to improvement.

How does one describe, mathematically, to what extent the data varies around the average? The mathematical process begins with finding the difference between each value and that average. Subtracting a larger number from a smaller number will result in a negative difference. If the differences are added at this point, the sum will always be zero. (This will be an important factor in a moment.) Squaring each difference before adding eliminates the negative difference problem. The sum of the squares of the differences between the mean and each value is the result. Since the number of individual values makes a difference in the sum, divide the sum by the number of individual values, just as in computing the mean. Remember that each difference was squared. To get back to a number that realistically represents variation in the data, take the square root of the result.

One last detail must be added. To calculate the mean of the data set, divide the sum of the values by the number of values in the data set. Each value varies independent of the other values. In computing the standard deviation, all but the last value can vary independently. Since the sum of the "non-squared" differences is zero, knowing all but the last number allows calculation of the last number. Therefore, divide the difference between the square of the sum of the differences between the individual values and the mean by one less than the number of values in the data set. If the number of values (samples) is small, the "degrees of freedom" affects the result of the standard deviation calculation dramatically. Dividing any number by 19 rather than 20 will affect the result much more than dividing by 99 rather than 100. The formula for computing a standard deviation of a data set (also noted as "σ") is:

$$\sigma = \sqrt{\frac{\Sigma_{i=1toi}\left(Mean - X_i\right)^2}{n-1}}$$

Calculating the standard deviation using the formula shown above is seldom done as an individual or team tries to understand the variability of data. The formula is shown to give the reader an idea of the mathematics behind the concept. A preferred approach is to use a computer and a statistical software package. Simpler yet is the use of a statistical calculator. Statistical calculators are inexpensive and easy to use.

Constructing an XBar&R Control Chart

Constructing an XBar&R Chart is not difficult. The preliminary phase of the process is, of course, to select a measure that is important to customer satisfaction. Next, determine the sample size the operator will be take, and how often. This will depend on the frequency of the output (more frequent requires samples more often), how costly the measurement is (more costly sampling equals fewer samples taken, and taken less often), and how critical it is to monitor the process closely. Exhibit 2-3 outlines the step-by-step process for creating the XBar&R Chart. Figure 2-4 shows a completed XBar&R Chart.

The completed XBar&R chart will visually display the variation and central tendency of the process, based on the measure selected. The plot of the various sample averages and the ranges of the sample values will show the variation of the process. The line representing the average of the sample averages, or the grand average, shows the central tendency of the process, over time.

After an adequate number of periods, compute and record the upper and lower control limits for the averages and the ranges. The control limits represent the limits expected of the averages and ranges when the process is operating in control.

The accumulation of average values and ranges will form a normal distribution while the process is in control and not changing. This is an important aspect of Control Charts. If the process in not in control, an abnormal distribution of values will be the result.

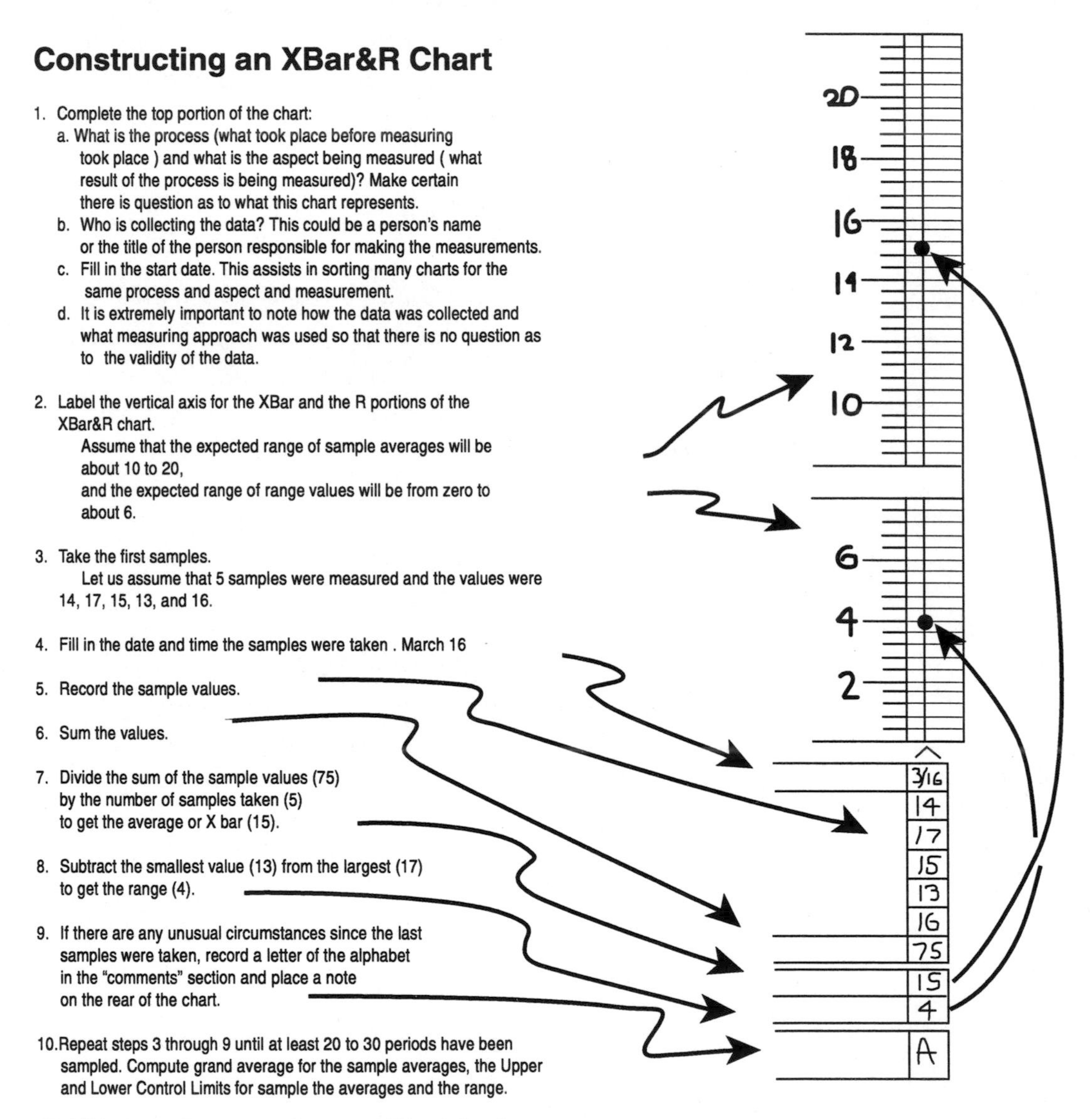

Constructing an XBar&R Chart
1. Complete the top portion of the chart:
a. What is the process (what took place before measuring took place) and what is the aspect being measured (what result of the process is being measured)? Make certain there is question as to what this chart represents.
b. Who is collecting the data? This could be a person's name or the title of the person responsible for making the measurements.
c. Fill in the start date. This assists in sorting many charts for the same process and aspect and measurement.
d. It is extremely important to note how the data was collected and what measuring approach was used so that there is no question as to the validity of the data.
2. Label the vertical axis for the XBar and the R portions of the XBar&R chart.
Assume that the expected range of sample averages will be about 10 to 20,
and the expected range of range values will be from zero to about 6.
3. Take the first samples.
Let us assume that 5 samples were measured and the values were 14, 17, 15, 13, and 16.
4. Fill in the date and time the samples were taken . March 16
5. Record the sample values.
6. Sum the values.
7. Divide the sum of the sample values (75) by the number of samples taken (5) to get the average or X bar (15).
8. Subtract the smallest value (13) from the largest (17) to get the range (4).
9. If there are any unusual circumstances since the last samples were taken, record a letter of the alphabet in the "comments" section and place a note on the rear of the chart.
10.Repeat steps 3 through 9 until at least 20 to 30 periods have been sampled. Compute grand average for the sample averages, the Upper and Lower Control Limits for sample the averages and the range.
20
18
16
14
12
10
6
4
2
3/16
14
17
15
13
16
75
15
4
A

Exhibit 2-3 Constructing an XBar&R chart.

XBar&R Charts

Process and Control Limit Computations

Estimating Sigma:

$$\hat{s} = \overline{R}/d_2$$

($\hat{\sigma}$ is "sigma hat" and is the estimate of the process standard deviation)

Computing the Process Limits:

$$UPL_{\bar{x}} = \overline{\overline{X}} + 3s$$

$$LPL_{\bar{x}} = \overline{\overline{X}} - 3s$$

(Where "s" is the standard deviation for the sample group averages. " $\hat{\sigma}$ " could substitute for "s")

Computing Control Limits:

$$UCL_{\bar{x}} = \overline{\overline{X}} + (A_2 * \overline{R})$$

$$LCL_{\bar{x}} = \overline{\overline{X}} - (A_2 * \overline{R})$$

$$\overline{R} = \frac{R_1 + R_2 + R_3 + \ldots + R_n}{n}$$

$$UCL_R = D_4 * \overline{R}$$

$$LCL_R = D_3 * \overline{R}$$

Constants for XBar&R Charts:

For lots (subgroups) of two:

$A_2 = 1.88$ $D_3 = N/A$ (The LCL_R will be zero for samples of two.)
$d_2 = 1.13$ $D_4 = 3.27$

For lots of three:

$A_2 = 1.02$ $D_3 = N/A$ (The LCL_R will be zero for samples of three.)
$d_2 = 1.69$ $D_4 = 2.57$

For lots of four:

$A_2 = 0.729$ $D_3 = N/A$ (The LCL_R will be zero for samples of four.)
$d_2 = 2.06$ $D_4 = 2.28$

For lots of five:

$A_2 = 0.577$ $D_3 = N/A$ (The LCL_R will be zero for samples of five.)
$d_2 = 2.33$ $D_4 = 2.11$

For lots of ten:

$A_2 = 0.308$ $D_3 = .233$
$d_2 = 3.08$ $D_4 = 1.78$

The above formulas and constants are thanks to Walter A. Shewhart, as popularized by his *Economic Control of Quality of Manufactured Product*, Van Nostrand 1931.

Exhibit 2-4 Process and control limits for XBar&R Charts.

Figure 2-4 charts the process that assists customers (installers) of "Analsoft 1.2" software. Five calls per day were chosen at random and timed. This chart, therefore, requires receiving five or more calls per day. If less than five calls had been received during any day, the person collecting the data would not have included that day.

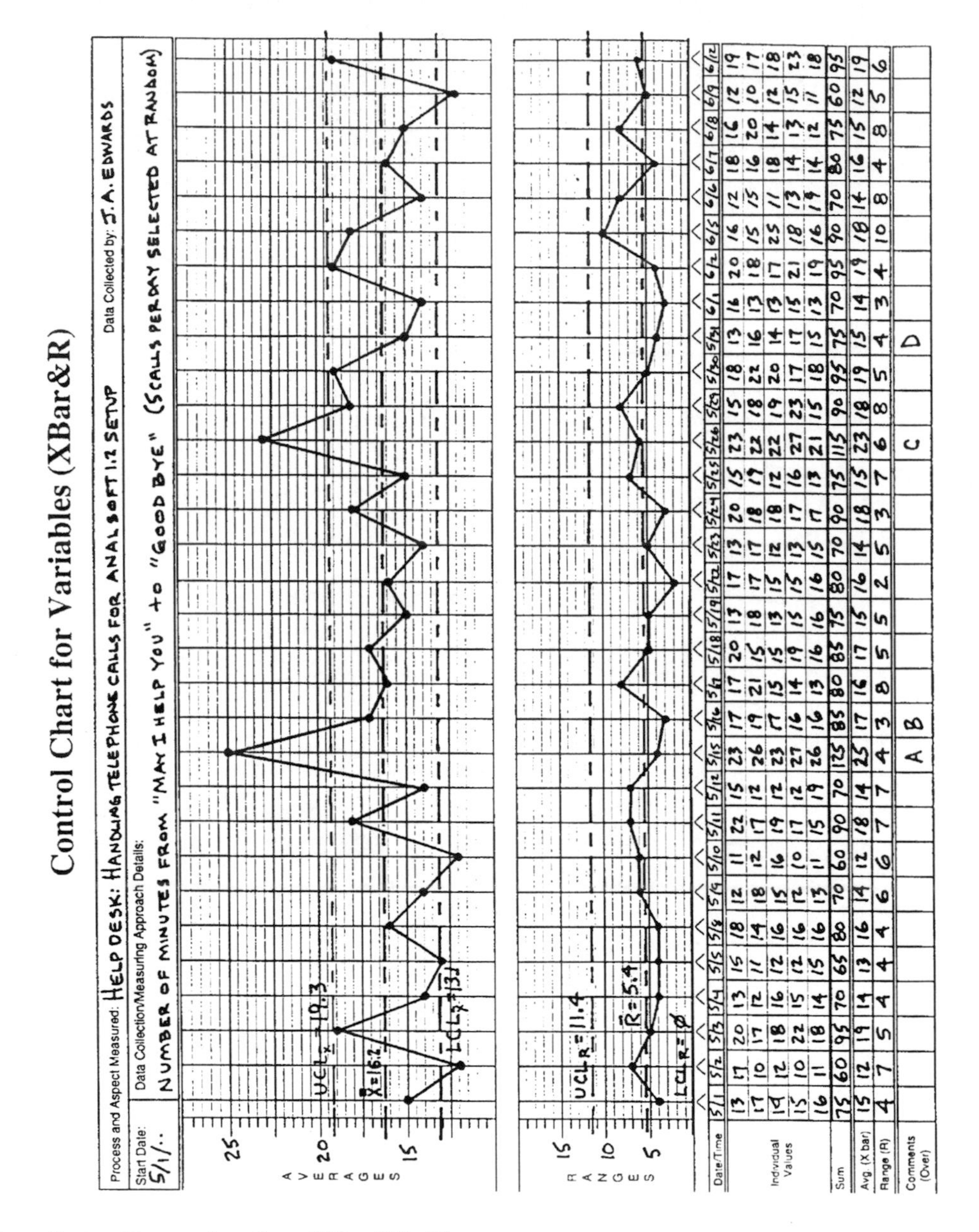

Figure 2-4 Example of an XBar&R Chart.

The average of the sample averages ($\overline{\overline{X}}$), the range average ($\overline{R}$), and the upper and lower control limits for the sample averages and ranges were computed using the formulas found in Exhibit 2-4. Lines representing these values have been added to the chart shown in Figure 2-4. The chart does not include upper and lower specification limits because management, industry standards or the customers did not specify limits. The customers do not have an exact expectation of the help process except to expect the help desk to take enough time to be of assistance and yet not take more time than is really necessary.

The Control Chart for the specific 31-day period clearly shows that the process sample averages are out of control. Five different days exceed the upper and lower control limits for the process. This strongly suggests that the time it will take for the help desk to assist an installer is not predictable. The grand average for the process ($\overline{\overline{X}}$) is 16.2 minutes. Yet the five-sample average for May 15 was 25 minutes. Remember, five sample values produced the sample average for May 15, some of which had to be greater than 25 minutes. The individual values portion of the Control Chart shows that the longest sample time was 27 minutes. Other indicators also suggest a process is out of control. A later section in this chapter, *Actions on Special Causes*, suggests actions to take based on how the sample averages or range values are behaving. These decision rules specifically apply to processes that are in control. The process operators must bring the help desk process into control, first.

The range values are in statistical control. Note that all range values are within the upper and lower control limits. The help desk personnel can predict that the range of sample times will range from zero to 11.4 minutes. They can use the decision rules shown in Exhibit 2-9 to suggest when the ranges are going out of control.

To understand this process better, compute an estimated sigma ($\hat{\sigma}$) for the sample averages. Compute the estimated sigma for XBar&R charts using this formula: $\overline{\overline{X}}$ divided by d_2 equals $\hat{\sigma}$. This formula and the value for d_2 are found in Exhibit 2-4. The computed value for the estimated sigma is 2.32. Dividing 5.4 (the range average) by 2.33 (the value for d_2 when five samples are taken) determines this value. Statistically, the range of values from three sigma less than the grand average to three sigma above than the grand average will include 99.74% of the sample averages. The estimated upper process limit is 16.2 (the grand average) plus three times 2.32 (the estimated sigma). The estimated upper process limit is 23.2. The estimated lower process limit is 16.2 (the grand average) minus three times 2.32 (the estimated sigma). The estimated lower process limit is 9.24. The upper and lower process limits are "estimated" because the values are based on the estimated sigma.

The help desk people can estimate that they can help a customer in no less than 9.24 minutes and will finish by 23.2 minutes. However, since the process is not in control, any given help call may take less or more than these minimum or maximum times. Special causes of variation could occur at any time, without warning.

The next step in improving this process is to identify causes of variation due to special causes. Chapter Eleven discusses special causes of variation in further detail. The Control Chart provides information in a form that will assist in identifying special causes and will also warn process operators of special causes creeping into processes. The first suspect in the help desk chart is May 15. The sample average is beyond the upper control limit and a note refers to an unusual occurrence that day. These two pieces of information may be vital. The fact that the sample average fell within the control limits on May 16 and that the person keeping the chart included another comment, may reinforce suspicions about May 15, or may offer additional information. This is why recognizing special circumstances and noting comments on the reverse side of the chart is critical.

X&MR Charts
Control Chart for Individual Measurements

Often, taking more than one sample at a time is not possible. This may be due to the limited output from the process, measurement difficulty, measurement cost or the requirement for destructive testing to get a measurement. If any of these factors are present, then use the Individuals Chart. The Individuals Chart is quite similar to the XBar&R Chart except that the "Averages" portion ($\overline{X}$) becomes the "Individual" (X) measurement portion of the chart. In addition, subtracting the most recent measurement from the previous measurement produces the moving range (MR) value, discarding a minus sign.

Calculation of the control limits is also similar to the XBar&R chart. Normally, most X&MR charts use only the previous period's value for the range calculation. However, using two more previous periods' values is possible. Using two or more previous periods to compute range will produce a range graph with more variation. The advantage of using three or more periods is the possibility of better understanding of how the process is behaving. Calculation of the control limits is also similar to the XBar&R chart. D_3 is not used since the lower control limit for the moving ranges is zero. Exhibit 2-5 shows the formulas and constants for X&MR Charts.

Control Limit Formulas for Individuals Charts:

$$UCL_X = \overline{X} + (E_2 * \overline{R})$$

$$LCL_X = \overline{X} - (E_2 * \overline{R})$$

$$\overline{R} = \frac{R_1 + R_2 + R_3 + ...R_n}{n}$$

$$UCL_R = D_4 * \overline{R}$$

$$LCL_R = 0$$

Number of samples used to compute the range, including current sample

	2	3	4	5	6
E_2	2.66	1.77	1.46	1.29	1.18
D_4	3.27	2.57	2.28	2.11	2.00

Exhibit 2-5 Formulas and constants used for computing an X&MR Chart UCL and LCL.

Constructing an Individual Samples Chart X&MR Charts

Constructing a Control Chart for Individual Samples is very similar to the construction of a XBar&R Chart. The differences are:

1. One takes one sample during each period or from each lot,
2. The difference between the current period value and the previous period's value determines the range, and
3. The formulas for computing the $LCL_{\overline{X}}$, $UCL_{\overline{X}}$ and $UCL_{\overline{R}}$ are different.

Remember. It is important that a person use the X&MR Chart only when the XBar&R Chart is not practical.

Figure 2-6 is an example of an X&MR Chart of the difference between estimated baggage weights and actual baggage weights. A baggage handler could use this X&MR Chart to understand the accuracy of the checked baggage weight estimating process. Since weighing every piece of baggage is time-consuming and could cause flight delays, the checker samples only one flight a day. Though an X&MR chart is not as "rich" in data as an XBar&R chart, the example X&MR chart does provide a picture of the weight estimating process.

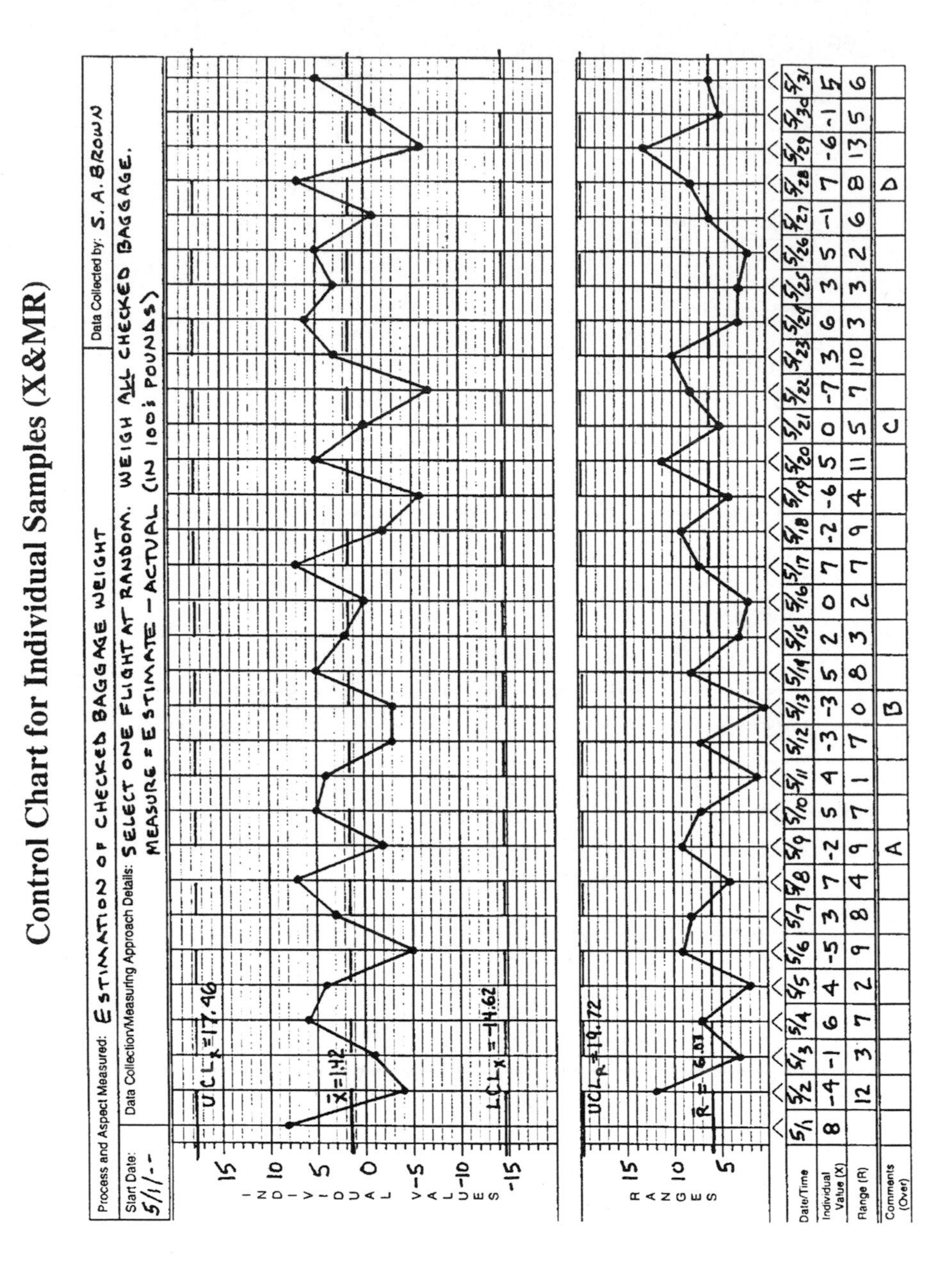

Figure 2-6 Example of an X&MR Chart.

Note in Figure 2-6 that the upper and lower control limits for the sample values are well beyond the greatest overweight and greatest underweight values. This would suggest that overweight as much as 1,746 pounds (17.46 hundred pounds) or underweight as much as 1,462 pounds (14.62 hundred pounds) could occur. This may or may not be acceptable for the type aircraft in this study. It is up to someone with expertise in aircraft and baggage weights to set an upper and lower specification for the estimating process. Those involved in the process can then begin the improvement of the weight estimating process to meet, then exceed, the specifications.

Control Charts for Attributes

Unlike variables, attributes are not strictly "measurable" in the way we normally think of measuring something. Instead, each attribute measurement results in one of two conclusions, each the opposite of the other. The result is either good or bad, right or not, defective or not. The aspect being measured makes it or does not make it. It passes or does not pass. Nothing is between the two extremes; a gray area does not exist nor does a degree of goodness or badness. Organizations often use no/no go gauges or strict standards to make attribute decisions. A typographical error is a defect, despite the consequence it might produce. A late report is a defect report, even if only a minute late. An unhappy customer is a defect, even if only a little unhappy. Use a scaled measurement (perhaps a one through 5 degrees of negative consequence) or discrete measurement (cost to correct the consequence) if the consequence that might result is critical.

Sometimes an output could have more than one type defect or more than one of the same type defect. One customer complaining about three problems is either an unhappy customer (the attribute is customer satisfaction) or it is three complaints to be resolved and prevented from occurring again (the attribute is the individual complaint). A customer with two complaints and a customer with four complaints is either two unhappy customers, six distinct complaints or an average of three complaints per customer. A builder might purchase 100 pieces of framing lumber. If five have knots making them unusable, the result is five defectives or a 5 percent defect rate. It is irrelevant that two of those five have two knots each, they are equally unusable. When cutting a piece of lumber to size, the result is either the right size or the wrong size. The person wanting the lumber cut could ask that the cut must be within one-eight of an inch of the specified dimension. Any measurement within that range produces a conforming cut. More than an eighth-inch too small or an eighth-inch too large is a defective cut. It is immaterial that the larger cut could be cut again to arrive at a conforming measurement. If a new refrigerator has a dent and a piece of decorative trim missing, it is a defective refrigerator and not fit for sale as a quality refrigerator. For the purposes of correcting defects and subsequent problem-solving activities, it is a refrigerator with two defects.

With four types of Attribute Charts available, deciding which of the four to use in a given situation is the first decision to make. The first decision is whether only one defect could occur or, even if more could occur, the sample is defective, just the same. The second decision is to consider the consistency of the process output quantity. If it is consistent over time, examine constant sample lot sizes. If the output quantity varies, then the lot sizes will probably vary. When the process output quantity is large and varies considerably, a constant sample size can be drawn from each period. These two criteria form an Attribute Chart decision matrix, shown in Exhibit 2-6. Chapter twelve presents other uses of the decision matrix.

Make one last decision before actually collecting the data and completing the chart: How many samples will be drawn from each period or lot? Always weigh the cost of taking samples and making measurements against the value of the Attribute Chart. Destructive testing of an expensive output requires weighing the additional cost of losing one sample against the value of the Attributes Chart. During the process of bringing the process under control, take more samples to be certain that the samples actually represent the process during that period. Once the process is under control, one sample may adequately represent the period prior to the sample. Use nondestructive measurement to confirm that all samples in the lot seem identical and the process remains under control. Then, only one destructive measurement might be appropriate.

	Varying sample size	Constant sample size
Only one defect per sample	p	np
More than one defect per sample	u	c

Exhibit 2-6 Choosing the right Attribute Chart.

During the period the process is out of control, ten samples from smaller lots will provide considerable information about the process, for that measure. After bringing the process into a state of control, reduce the sample size to five. Ten and five are chosen to reduce the mathematics required to complete the chart. If the lot sizes are larger or the periods produce larger quantities (perhaps more than 5,000), sample sizes of fifty or one-hundred might provide more reliable information. Again, consider the economics of retrieving the samples and making the measurements.

Make and record the measurements on the Attribute Chart. Record the sample size, total number of defects in all the samples and the number of defective samples. As in other Control Chart types, not tampering with the process while bringing the process into control is critical. After 20 to 30 periods or lots, the Attribute Chart can be completed and analyzed.

Exhibit 2-7 shows how to use each of the four Attribute Chart types to provide data on the process of getting automobiles to a dealer without dents. The chart notes the results of the analysis and sample size to take from future sample lots.

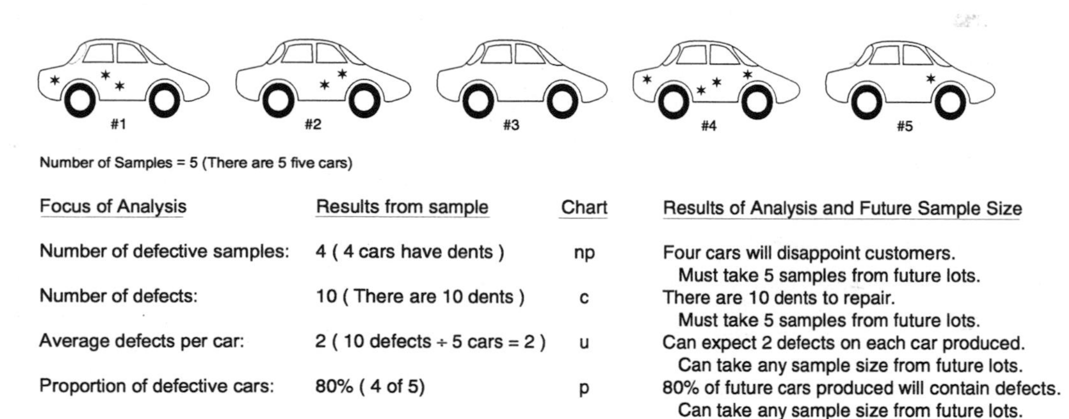

Focus of Analysis	Results from sample	Chart	Results of Analysis and Future Sample Size
Number of defective samples:	4 (4 cars have dents)	np	Four cars will disappoint customers. Must take 5 samples from future lots.
Number of defects:	10 (There are 10 dents)	c	There are 10 dents to repair. Must take 5 samples from future lots.
Average defects per car:	2 (10 defects ÷ 5 cars = 2)	u	Can expect 2 defects on each car produced. Can take any sample size from future lots.
Proportion of defective cars:	80% (4 of 5)	p	80% of future cars produced will contain defects. Can take any sample size from future lots.

Exhibit 2-7 Example of Attribute Chart possibilities.

Constructing an Attributes Chart

Begin constructing an Attribute Chart by completing the top portion of the chart: Check the appropriate box at the top of the chart, describe the process and aspect being measured, who is collecting the data, the start date of the chart and the data collection methods. At this point, fill in the scale values. Lacking an understanding of how the data will behave, gather the data first, then fill in the values for the scale later.

For each period the data is collected, record the:

1. Date and/or time the output was produced.
2. Sample lot size. This need be done only once if a c or np Chart.
3. One of the following:
   Number of sample cases with defects (np Chart), or
   - Total number of defects in the sample lot (c Chart), or
   - Average defects per sample or case in the lot (u Chart), or
   - Percent of defective samples or cases (p Chart).
4. Comments. Make certain to note any unusual circumstances surrounding the sample period. This helps in analyzing the chart for special causes.
5. Translate the np, c, u, or p value to the graph portion of the chart.

Continue this process for 20 to 30 periods. As stated earlier in this chapter, do not tamper with the process during this period.

After the 20 to 30 periods, calculate the average for the np, c, u or p values. Add the average line to the graph portion of the chart. Next, compute the upper and lower control limits for the chart using the appropriate formula from Exhibit 2-8.

Specifications are not a part of the Attributes Chart at this point as they were for Variables Charts. The specifications for the charted output measure are part of the attribute measure. However, goals for reduction of defect levels can be added to the chart. Customer expectations, competitive pressures or the need to reduce the cost attributed to the cost of non-quality drive these goals.

Figure 2-7 shows a completed Attributes Chart. Note the difference between it and the XBar&R and X&MR Charts. Though the Attribute Chart does produce a "picture" of the process, the calculations are simpler and the data points are easier to record on the graph. However, the chart does not present as much as the other charts. Variability within each sample lot (range) is not present.

p Charts

$$LCL_p = \overline{p} - 3\sqrt{\frac{\overline{p} \times (100 - \overline{p})}{n}}$$

$$LCL_p = \overline{p} - 3\sqrt{\frac{\overline{p} \times (100 - \overline{p})}{n}}$$

u Charts

$$UCL_u = \overline{u} + 3\sqrt{\frac{\overline{u} \times (1 - \overline{u})}{n}}$$

$$LCL_u = \overline{u} - 3\sqrt{\frac{\overline{u} \times (1 - \overline{u})}{n}}$$

np Charts

$$UCL_{np} = \overline{np} + 3\sqrt{\overline{np} \times (1 - \overline{np} / n)}$$

$$LCL_{np} = \overline{np} - 3\sqrt{\overline{np} \times (1 - \overline{np} / n)}$$

c Charts

$$UCL_c = \overline{c} = 3\sqrt{\overline{c}}$$

$$LCL_c = \overline{c} - 3\sqrt{\overline{c}}$$

Exhibit 2-8 Formulas for Attribute Chart upper and lower control limits.

Control Chart for Attributes

☐p ☒np ☐c ☐u

Process and Aspect Measured: RETURNING BOOKS TO CORRECT SHELF LOCATION

Data Collected by: J. R. MARTINEZ

Start Date: 5/1/.. WEEKLY

Data Collection Methods: CHOOSE DAY OF WEEK TO CHECK AT RANDOM MONDAY A.M. SAMPLE 100 BOOKS SHELVED BEFORE OPENING LIBRARY. BOOK NOT SHELVED CORRECTLY EQUALS DEFECTIVE SAMPLE

NUMBER OF BOOKS (OF 100) MISSHELVED

18 16 14 12 10 8 6 4 2

UCL_{np} = 10.4

$\bar{p}$ = 4.52

LCL_{np} = −1.37 (THEREFORE = Ø)

Date/Time	5/1	5/8	5/11	5/27	5/30	6/11	6/14	6/22	6/26	7/8	7/11	7/19	7/30	7/31	8/12	8/17
Sample Lot Size	100 EACH PERIOD →															
np # of Samples With Defects	5	4	5	5	6	4	7	5	3	4	4	5	8	5	4	4
c # of Defects																
u Avg. Defects (Sample Lot)																
p % Defects (Sample Lot)																
Comments (Over)													A			

Date/Time	8/22	8/30	9/4	9/16	9/21	10/1	10/5	10/13	10/20	10/24	10/30	11/7	11/19	11/21	11/29
Sample Lot Size															
np # of Samples With Defects	5	3	4	4	5	3	6	3	4	5	2	4	5	7	2
c # of Defects															
u Avg. Defects (Sample Lot)															
p % Defects (Sample Lot)															
Comments (Over)						B					C			D	

Figure 2-7 Example of an Attribute Chart

Actions on Special Causes

Having a set of decision rules will govern decisions to watch a process more closely, take corrective action, or shut down the process until identification and correction of the cause of the problem. If life is at stake, a defect in a good or service would affect a critical customer, or the defect would be expensive to fix, change the "concern" rule to "immediate action" and the immediate action rule to "shutdown." Exhibit 2-9 outlines the decision rules.

The decision rules are statistically valid. Though producing four or more points alternating above and below the average is statistically possible for a process (sample averages), it is statistically unlikely that this situation could occur naturally. The process operator should them pay particular attention to the measures of the process. Think about what could have changed in the process. The detailed process analysis form shown in Chapter 3 will help the operator in thinking about all the factors that could bring in a special cause of variation. However, if a series of seven points fall above the process average, it is highly unlikely that random variation could be the cause. Assume that something special has moved the process average to a higher level. If a process average shift would cause critical problems, then shutting the process down would become the immediate action. Again, the detailed process analysis form in Chapter 3 will become a valuable tool.

Decision Rules for Control Charts

Series suggesting the need for concern:

- A series of 4 or more points alternating above and below the average.
- A series of 7 or more points evenly distributed between the upper and lower control limits.

Series requiring immediate action:

- A point outside the control limits.
- A series of 7 points above or below the average.
- A series of 7 points rising or falling.
- A series not appearing to be random.

Series requiring process shutdown and correction:

- A series of 5 points near the UCL or LCL.
  (Within the plus or minus third sigma area.)
- A series of 2 points outside the control limits.
  (Within the plus or minus fourth sigma area.)
- A point way outside the control limits.
  (More than a one-sixth control limit span beyond the upper or lower control limit, which would place the point more than four sigma beyond the average.)

Exhibit 2-9 Decision rules for Control Charts.

Single Sample Lots and XBar&R Charts

Variables Control Charts for averages pose a special problem. The two values used to compute the control limits are the average of the averages and the range of the values. Both carry enough information to present a picture of the health of a process. However, it is possible to get sample values beyond the process control limits or the limits of the specifications of the process. The average of the samples might not show a problem. The range should, but may not.

For example, consider a process with a grand average of 100, an average range of 10, the upper control limit at 105.77 and the lower control limit at 94.23. The process operator has just measured five samples: 104, 103, 105, 112 and 103. The average of this sample lot is 105.4, and the range is nine. The average and range appear to be within the established limits. But, what about the sample that measured 112? It is outside the control limits. Should the operator stop the process or should the operator attribute this sample to normal variation that could occur once in a great while?

The example points out the need for an additional set of decision rules. Sometimes it may become necessary to decide to stop a process based on the individualsample values rather than the sample lot average or range.

Figure 2-9 provides a simple set of decision rules for shutting down a process based on a single sample lot. The process operator should take the samples and record the individual values. One-hundred percent of the values in the two to three sigma ranges suggest a process shutdown. Consider the plus two to plus three range and the minus two to minus three range, combined. Forty percent of the values in the three to four sigma ranges suggest a shutdown. Any value beyond the four sigma points requires a process shutdown.

These rules have less statistical validity than the decision rules for sample averages presented on the previous page, but will nevertheless provide guidance to a process operator. The operator must also factor in the critical nature of the output of the process. Will the problem become hidden until much later? Will correction of the problem downstream be expensive? Is life at stake? Will a critical customer become disappointed due to a problem in this process? A "yes" answer to any of these questions suggests a need for a process shutdown based on the single sample lot decision rule.

A single point higher than 4+ sigma: Shutdown required!

+4 sigma

40% of the points in this area, and 3- to 4- area: Shutdown suggested

+3 sigma(UCL)

100% of the points within the +2 to +4, and -2 to -4 sigma areas: Shutdown suggested

+2 sigma

+1 sigma

Average (Mean)

-1 sigma

-2 sigma

100% of the points within the -2 ro -4 and +2 to +4 sigma areas: Shutdown suggested

-3 sigma (LCL)

40% of the points in this area, and the +3 to +4 area: Shutdown suggested

-4 sigma

A single point lower than 4- sigma: Shutdown required!

Figure 2-9 Decision rules for a process shutdown suggested by single sample lotss.

Customer data (Chapter 1) and process data (this chapter) arms an individual or team for the corrective actions or improvement activities in Chapter 3 through 5.

Chapter 3

Process Documentation and Analysis

Processes

Ideally, processes add value to what comes into the process, providing something of value to customers of the process. Supplies (materials, services and information) come into the process from the process suppliers. A person (or a group of people) does something with the inputs, to deliver an output to the next process or the customer. If the output meets the expectations of the next process or customer, the process is producing a quality output. If not, process analysis must take place to improve the process.

A process could range from the less complicated, such as keying a handwritten sales order into a computerized order entry system, to the more complicated, such as building a new office structure. Complex processes are difficult to analyze for quality and productivity improvement because they are most often a collection of many interconnected processes. Process analysis should always be done on the smallest portion of a complex process. Rather than analyze the process of building an office, the person or team should focus on a piece such as putting in the foundation. However, "putting in the foundation" may not provide enough focus when one considers that the foundation process includes determining where the foundation should go, digging, erecting forms, placing reinforcing rods, pouring concrete, and removing forms. The focus should be on the process that is most critical to success or the process most likely to provide a major improvement. The key is to focus as much as possible for the greatest opportunity.

Flow Charts

A Flow Chart is a simple, logical sequence of events that must (or should) take place to produce a distinct output. Flow Charts often picture a collection of processes, and could even include the steps within each process. Thus, a team could chart the entire employment process from the decision to hire a new employee to the point a new employee is on the job, or the team could chart the steps required to obtain a signed requisition for a new employee. One could chart the entire process of ordering, receiving and stocking the item, or one could chart the process of stocking a specific item in the right place. Charting a collection of processes could include so many parts or steps that the chart occupies an entire wall of a meeting room, and some often do. If the chart is massive, the next step should be to break it into manageable pieces.

Charting a macro process such as hiring a new employee takes a great deal of time and requires a team to produce a complete chart. The chart of a macro process clearly shows the mass of the process and why the process consumes so much time and energy. However, charting a more confined process is much easier. Consider the process of routing a telephone call requesting service.

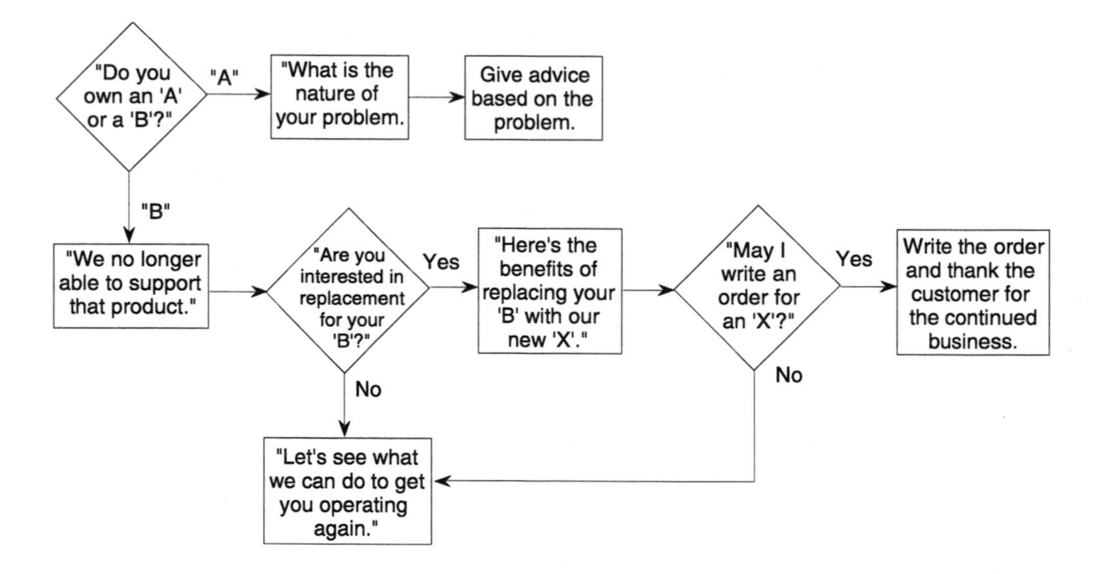

Figure 3-1 Simple Flow Chart of customer service representative taking a telephone call requesting assistance.

Flow Charts provide opportunities to identify bottlenecks, inappropriate steps or non-value added activities. Flow Charts also afford opportunities to improve product or service quality by rearranging the flow or combining steps to improve efficiency. Then, the person or team charting the process can identify and apply measures within the process, at appropriate points. The Flow Chart is also useful when training new process operators and is a visual documentation method as processes change in step with changing requirements of the process.

Flow Charting begins with a clear picture of the current process or system. The use of 3M brand Post-it® notes, or the equivalent, provides an excellent method for constructing a Flow Chart. Construct the Flow Chart on a flip chart, white board or wall. Post-it® notes permit easy modification of the chart during construction to represent the current situation accurately. Then, modification of the chart for improvement is easy. An alternative is to use note cards on a table or taped to a vertical surface.

The 3" by 5" rectangular Post-it® notes or note cards are a suitable size for process steps. Tearing the corners off a rectangular note represents the initiation of a process. A 3" square Post-it® note, rotated 45 degrees, represents a decision point. A decision step usually produces two alternatives. Often the alternatives are simply "yes" or "no." Other decision steps could produce three or more options. These options could relate to specific processes, process steps, actions, directions, priorities or locations. Overlapping two or more square notes will provide four or more decision options. End steps are circles. Tear or cut the corners off a square note to show an end step. Represent a filing, record keeping or storage step by tearing an "S" shape along the bottom of a rectangular note. Additionally, the note card color, or the color marker used, could represent how critical the step is within the process. Pink represents a critical step, yellow a key step, green a process step and blue an optional step. Figure 3-2 shows the common Flow Chart shapes.

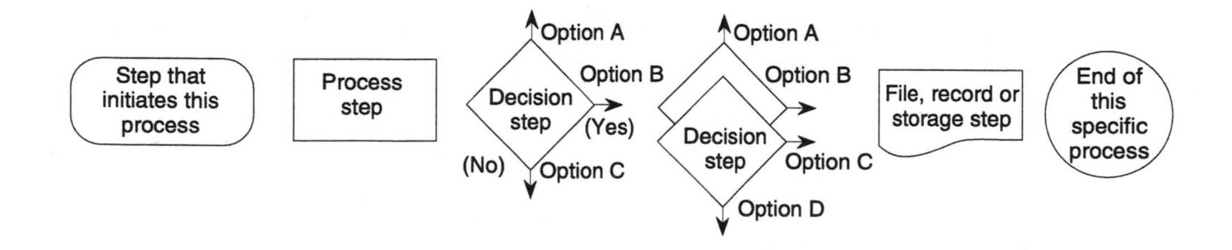

Figure 3-2 Five common symbols used for Flow Charting a process.

The simple Flow Chart shown in Figure 3-1 could become more complex as the group thinks about how service calls are actually handled rather than how the person should handle the calls. To add detail to the chart, team members should put themselves in the caller's place or actually survey typical callers to learn what really happens. This will also help in identifying the key or critical steps.

Some steps may be processes worthy of their own Flow Charts. For example, what information should the service representative obtain from a caller requesting service? Perhaps asking a customer questions to complete a service request form is a separate process, justifying its own Flow Chart.

The more a team works on a chart the more complex it becomes. What one might think of as a simple process is often a complex process or a collection of processes. If the process being charted is a collection of processes, then it is time to break it down to its component processes. Many cases exist where a process flow chart covers one entire meeting room wall. The team should start with the process that is most critical or is the source of critical problems. If the charted process is a single, but highly complex process, the improvement initiative becomes to break it into several more manageable processes or to simplify the process, dramatically.

Top-Down Flow Charts

The Top-Down Flow Chart is a useful variation of Flow Charting. Use a Top-Down upon discovering that a linear group of primary steps forms the basis of the process, but that each step consists of two or more supporting steps. Block the primary steps in the process across the top of the page or display board. Beneath each major block are the supporting steps, or processes, necessary to complete that block. It is sometimes easier to focus on each primary step rather than the entire process. The team could then focus on the components of the primary step with bottlenecks, delays, problems, variation, high cost, non-value added or complexity. Flow Charting the Eight-Step Problem-Solving and Improvement Process (Figure 3-3) is an example of an appropriate use of the Top-down Flow Chart. Eight primary steps make up the process. The first step, project selection or assignment, initiates the process. The other primary steps include the steps necessary to complete each. The chart makes it clear what must take place in each primary step to complete the step. Systematic Problem Solving is discussed further in Chapter 10.

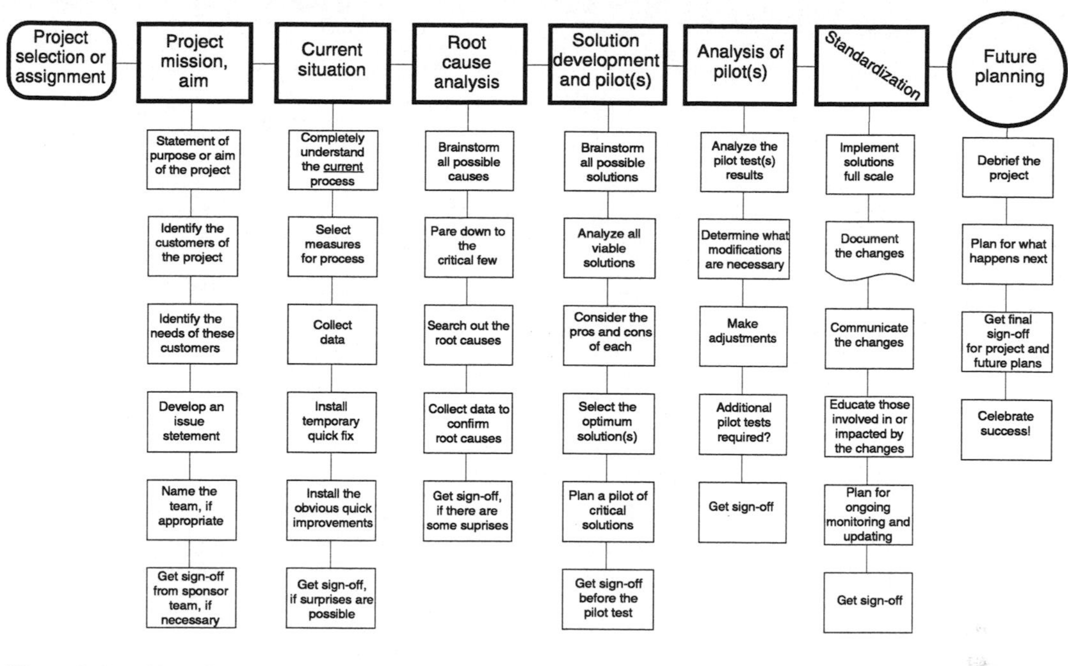

Figure 3-3 Top-Down Flow Chart of a systematic problem-solving process.

Add the concepts of the Top-Down Flow-Chart to a standard Flow Chart to produce a highly flexible Flow Chart. Figure 3-4 shows an example of this approach.

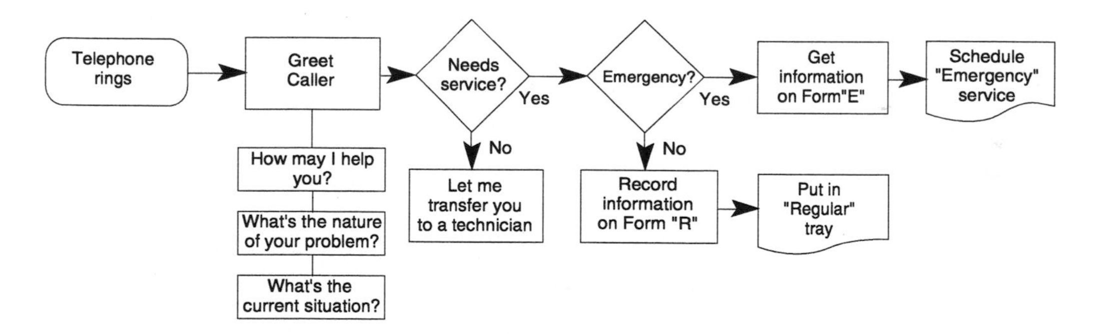

Figure 3-4 Simple Flow Chart with one block expanded utilizing a Top-Down approach.

Work-Flow Diagrams

Work-Flow Diagrams visually represent the physical sequence of work activities that must take place to complete a process, or collection of processes. The diagram also shows the connections between activities that are taking place in parallel. Work-Flow Diagrams will help a team or individual better understand a current process and the opportunities to improvement. The Work-Flow Diagram of the improved process will show the improvement visually. A Work-Flow Diagram is particularly useful in removing complexity, redundancy and non-value added activities from a process.

The first step in creating a Work-Flow Diagram is to produce a diagram of the work area, or work areas involved, on a large piece of paper or a white board. Diagraming a single room, several rooms, a building or distant geographic locations is possible. Therefore, the full Work-Flow Diagram may consist of several work areas with any degree geographic separation.

Beginning with a work area diagram, the process operator(s) or an observer documents the movement of people, materials and information. It is best that the supervisor or manager of the processes does not itemize the processes. The process operators know what activities must take place. An uninformed observer simply documents each activity as it takes place. An alternative would be for a team associated with the process to Brainstorm (see Chapter Nine) all activities involved in accomplishing the process and combine groups of minor activities if the list is quite long. Record each distinct activity, or collection of minor activities, on a Post-it® note. Start by using green notes, only. Later, someone can use other color notes to suggest special characteristics of certain activities. The notes are then placed on the flip chart or white board in the location where the activities occur. Arrows are drawn between activities to show the sequence of occurrence.

Some activities do occur in parallel, sometimes involving two or more different operators. For example, generation of a customer letter could be occurring while another activity is taking place generating a computer printout. Sometimes one activity must take place by an operator before another operator can continue an activity. An example would be that the person placing a customer letter and computer printout in a window envelope must have the letter and printout before that activity can take place.

If material or information moves from activity to another, show that movement on the chart. To some extent, the chart now represents a time exposure of what occurs over a specific time to accomplish something.

The Work-Flow Diagram presented in Figure 3-5 records the path that three employees traveled to make copies. One immediately recognizes that the layout of the copy/supply room is not very efficient, simply because the people using the room must move to so many places within the room. Rearranging the room would result in less foot traffic, fewer bottlenecks when several people are in the room and fewer chançes for items to be lost. The result would be an improvement in quality and productivity, and higher employee satisfaction.

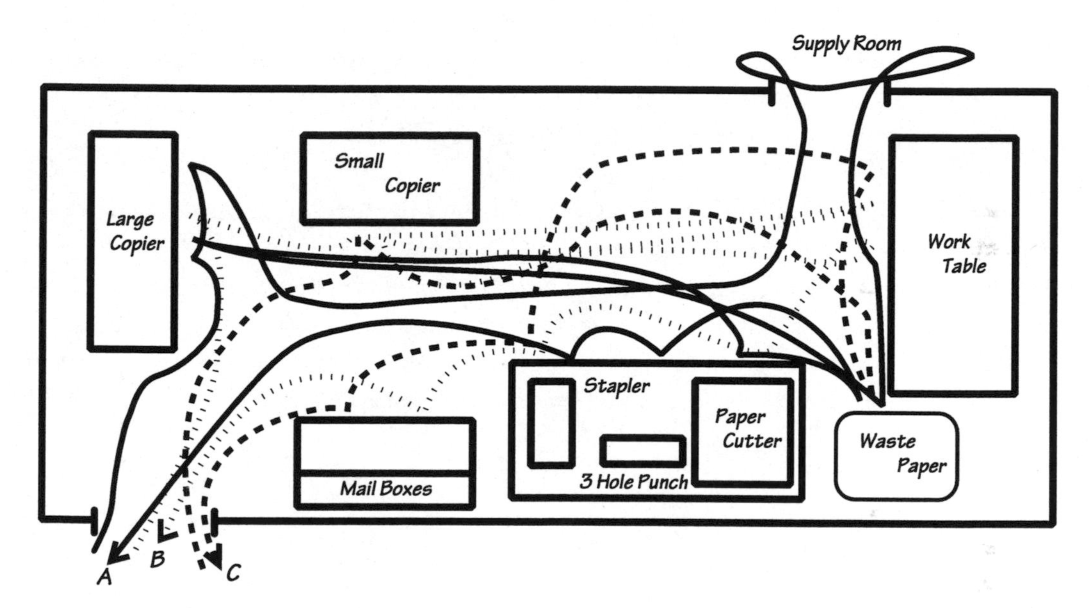

Figure 3-5 Example of a Work-Flow Diagram showing the paths of three typical users.

Activity Diagrams

Activity Diagrams are a special type of flow chart, showing the variety of activities that must take place to accomplish a task or complete a process. The Activity Diagram also includes the sequence of activities and the direction of work flow. The diagram is useful when analyzing a process to isolate non-value added activities, bottlenecks, needless complexity, unnecessary delays, and inappropriate or counterproductive activities.

Figure 3-6 is the Activity Diagram for scanning a customer's purchases at a supermarket. Note how some activities form a loop, such as scanning each item in sequence and rescanning items a second or third time. Other activities occur in one of several possible routes depending on the answer to a key question. For example, does an item require special handling or bagging after scanning? The scanning process ends upon entering the last item. This initiates the "coupon" process and then the "payment" process. While these two processes are occurring, the "bagging" process takes place.

After diagraming the scanning process, the problem solving and improvement activities can take place. Is it necessary to scan each individual item or is it more efficient to collect similar items a first step? Is arranging the items by category a beneficial activity? Can anything be done to reduce or eliminate "system does not register the item properly" problems? What, exactly, constitutes "item requires special handling or special bagging"? Is it possible to transfer this decision process to the bagger?

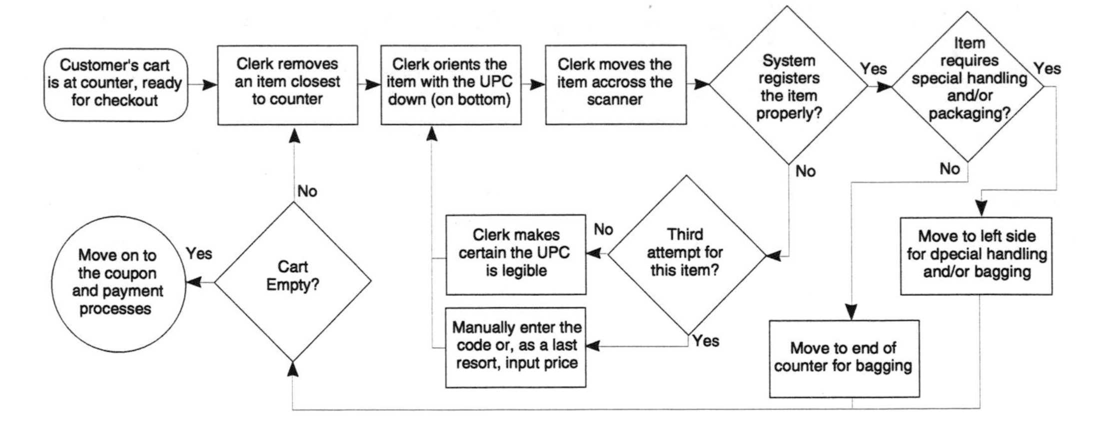

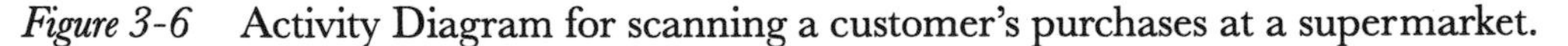
Figure 3-6 Activity Diagram for scanning a customer's purchases at a supermarket.

Completing the Activity Diagram of the current process before attempting to improve the process is important. The comparison between the old process and the new provides a measure of how much improvement took place. Exhibit 3-1 outlines the procedure for creating an Activity Diagram.

The Activity Diagram provides one additional benefit to organizations, making it an even more valuable tool. The diagram should become one of the new employee job training tools. Activity Diagrams are the visual equivalents of job process documents.

Steps for creating and using an Activity Diagram

1. Document all the activities that make up the process, noting:
   a. What occurs,
   b. Time required to complete (or, if variable, mean and range of time),
   c. Prior activities required, and
   d. Inputs required.
   Record each activity on a separate note card or Post-it(R) note.
2. Determine what initiates the activity diagram (the process) and at what concludes the process with an obvious, observable, measurable output.
3. On a large work surface, flip chart or white board, arrange the cards to reflect the current process. This is not the time to improve the process. If ideas surface for improvement, note them and set aside for Step 5.
4. Add up the times representing the longest sequence of activities required to complete the process. Draw (or use cards to note) a time line along the bottom of the chart. Align the cards to show which activities occur at what time. Two or more blank cards can be added to the right of activity cards with longer times to complete. This provides a visual clue of activity times. Record or document the current process for a baseline to compare the improved process.
5. Now is the time to begin improving the process. Look for:
   - Time-consuming activities to shorten,
   - Activities with large ranges of time to reduce,
   - Redundant activities to eliminate,
   - Unnecessary loops,
   - Activities to combine,
   - Delays to reduce or eliminate,
   - Activities that could occur during necessary delays,
   - Activities that could run concurrently,
   - Activities that could benefit from staging by others better equipped to do them,
   - Activities simply not required,
   - Activities to rearrange to improve effectiveness, and
   - Critical, but missing activities.

Exhibit 3-1 Procedure for creating an Activity Diagram.

One reason that Activity Diagrams are not used more often to improve processes is the fact that they can become quite complex, quickly. For example, the process diagramed in Figure 3-7 is that of making two pieces of buttered toast. The upper diagram is the unimproved process. In its current state, the process has a built-in delay while the person toasting the bread stands and waits for the toast to pop up. The bottom diagram shows the improved process. Improvement reduced the process time by 25 seconds. The improvement was to move two activities to the period during which the operator waits for the toaster to finish. If making buttered toast is the operator's job eight hours a day, the 25 seconds would add up to many minutes. In fact, the operator will save about 42 minutes a day if the toasting and buttering operation takes place 100 times during the work day.

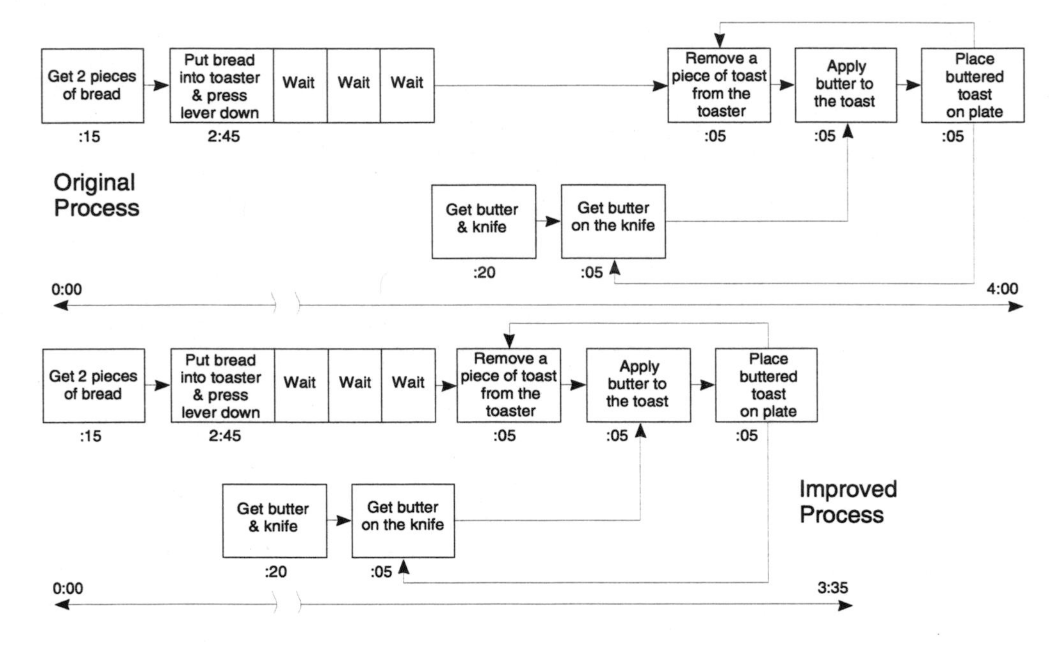

Figure 3-7 Before and after improvement Activity Diagrams of the process for creating two pieces of buttered toast.

S•I•P•O•C
Suppliers-Inputs-Process-Outputs-Customers

Much of the individual problem-solving and the multitude of incremental improvements come from a focus on processes. Hundreds, even thousands of processes make up an organization. Flow Charting shows linkages between processes and the steps within each process. Work-Flow Diagrams show how work activity flows within a process or a physical area of the organization. Activity Diagrams show the interconnections between processes or between steps within a process. Eventually it comes down to analyzing a process. S·I·P·O·C is the analysis of what inputs flow into the process, how the process transforms these inputs and what outputs flow out of the process to customers.

All processes consist of three major components:

- The output is what the process produces. This output can be an actual good, something physical which can be touched and physically measured, or a service, something that occurs. Examples of outputs are a handwritten memo, a machined part, a finished product, an insurance policy, a mowed lawn, a haircut and a hiring decision.
- The inputs are goods, services and information that flow into the process. The process consumes or modifies these inputs, or allows the inputs to flow through without modification. Examples of inputs are specific information, memo paper, unfinished metal parts, product components, an insurance form, long grass or long hair and screened candidates for a job. The process does something with these inputs.
- The process itself consists of everything else necessary for the process to operate at a level of quality consistent with the requirements of the output. The four main components of the process itself are:
  1. The standard of quality at which the process intends to operate as determined by the organization, the process operator, or the customer of the process,
  2. The laws, regulations, industry standards, policies, procedures and general instructions that set up the boundaries for the process,
  3. The facilities, equipment and tools necessary to produce the output to the quality level demanded by the standard of quality for the process, and
  4. The capabilities of the process operator as determined by the operator's knowledge, skills, experience and physical attributes.

The person (or persons) in immediate control of a process is the process manager (if given more responsibility and authority for the process) or the process operator (if given less responsibility and authority for the process). Processes receive goods, services and/or information from suppliers to the process. Then process produces (adds value to) the goods, services and/or information to meet the requirements of the process customer. The process consumes some goods, services and/or information and, therefore, are not parts of the output. Exhibit 3-1 offers a form for process analysis using the S·I·P·O·C approach.

S•I•P•O•C Analysis

Process: ______________________ Date: __________

Analysis by: ______________________

Suppliers:	Inputs:	Process:	Outputs:	Customers:

Areas of concern: ______________________

Exhibit 3-2 The S·I·P·O·C form.

Though three primary components make up a process, six steps are necessary to analyze a process for improvement completely:

1. Determine exactly what the customer requires of the output of the process. Identify the customer's exact needs and wants. (Reduce subjective measures and guessing.)
2. Consider the standards (level of quality) and values (concern for quality) that relate to the output of the process. What level of quality does the customer expect? What standards of quality does the process operator place on the process output?
3. Confirm the appropriate application of general instructions, policies and procedures, and laws and regulations.
4. Determine the extent to which the facilities, equipment and tools necessary to the performance of the process are appropriate and are available.
5. Does the person (or people) involved in the process have adequate experience, knowledge, skills and training to operate the process at the quality required? Is the person in a psychological environment conducive to optimum performance?
6. Have the appropriate inputs (that which the process passes through, modifies or consumes) been provided, as required?

The form shown in Exhibit 3-2 provides a simple method for determining what comes into a process and from which suppliers, what transformations take place and what comes out of the process, and who receives the output.

Complete the form by first focusing on the process being studied or improved. Make certain it is not a collection of processes. Next, identify the customers of the process. Begin with those internal processes which make use of the outputs of the process. These are the internal customers of the process. Then continue to the ultimate customer, but only if possible, practical and worthwhile. Move "backwards" to outputs required to meet the requirements of these customers, focusing on the immediate internal customer if subsequent processes will then meet the needs of external customers. After identifying all, or at least the critical, needs of the customers, move into the process and list those activities that must take place to produce conforming outputs. Move upstream to the suppliers. Identify each supplier, or at least the critical ones, if there are many. Next, move to the inputs required of the suppliers to operate the process properly. Remember. Inputs take the form of goods (materials, parts, power, etc.) and services (assistance, information, etc.).

Equipped with the S·I·P·O·C analysis, the team or individual can begin determining how to meet customer requirements, better. A search for inconsistencies between what someone designed the process to produce and what the customer needs should be the first effort. Then, the individual or team should go into the process to learn what improvements might improve outputs. Finally, the search can move on to the suppliers and what they supply. Are the supplies appropriate to the needs of the process and the customers of the process?

Detailed Process Analysis Form

Often, analyzing a process in greater detail is necessary. S·I·P·O·C is limited. The Detailed Process Analysis Form leads the analysis to a depth that should expose problems not exposed by a simple S·I·P·O·C analysis. This form directs the team or individual to examine the components of a process in detail. This leads to the identification of likely problems and opportunities for improvement. Figure 3-8 shows the basic form.

Processes consist of five distinct components. Standards the operator sets for the process make up the first component. Ideally, these standards are consistent with standards external to the process. Sometimes these standards of quality are set in absence of standards coming from the customers, regulations or the organization. The second process component is the quality standards set outside the process. Remember that external organizations often set quality standards. These organizations include various federal, state and local laws, regulations and ordinances. Industry standards also set quality standards. The third component is the physical part of the process. This includes the facility or work area, the equipment and the tools necessary for a process to operate properly. The fourth process component is the operator or operators. Operators bring knowledge, education, experience and skills to the process. This component could also include the operator's attitude, but attitude also affects the quality standards set by the operator. The fifth component consists of the inputs to the process. The process consumes, modifies or passes through inputs. Inputs modified by the process should have value added during the process.

Use of the Detailed Process Analysis Form is not complex. The operator or team begins with the output and fills in the blanks in sequence, finishing with a list of the inputs. As the name suggests, this form is similar to the S·I·P·O·C form, but adds detail. These details provide much more information for improving the process.

Use of the Detailed Process Analysis Form

As with the S·I·P·O·C analysis, the Detailed Process Analysis Form is completed from right to left. Begin with the requirements of the output or the process and progress left to the nature of the inputs. The information generated by the Detailed Process Analysis Form provides clues to what might be missing from the process, what might be causing problems, or what improvements could possibly produce a conforming output.

To produce conforming outputs, whether goods or services, first identify the customers of the process and what requirements they have of the outputs. This is the notion of beginning process improvement with a customer focus.

Structure a process in a way that transforms inputs into conforming outputs. Processes consist of four components:

1. The quality standards imposed on the process by the organization's environment and culture, and the process operator's standards. These standards may allow less-than-defect-free quality ("just get it close" or "quality must be 98%") or the operator's attitude may be "quality that just gets by." A supportive psychological environment must be associated with the process. A non-supportive environment could reduce the operator's commitment to customer satisfaction.
2. The combination of laws, regulations, standards, policies, procedures and specific instructions governing the process. These may interfere with the highest level of quality (improvement that becomes difficult to carry out because of regulations) or the specific instructions may not be appropriate to what is being done at a particular time.
3. The physical necessities of the process. These include the facilities (suitable enough as not to interfere with the process), equipment (suitable and functions properly) and tools (also must be suitable and function properly).
4. That which the process operator brings to the process. The operator must have the proper knowledge, skills, experience, and physical capabilities.

Also, necessary is a complete knowledge of the required inputs, along with an identification of those suppliers who can supply the necessary inputs. Knowledge of the suppliers often goes beyond the capabilities and authority of a process operator. This requires support of others in the organization for complete process analysis to be possible.

Detailed Process Analysis Form

Suppliers Provide **Inputs**

6a. That which is consumed by the process

6b. That which is modified by the process

6c. That which passes through the process without modification

Transformation Takes Place in the **Process**

3a. Laws and regulations applying to process or output

3b. Industry standards

3c. Organizational policies and procedures

3d. Specific instructions

2a. Quality standards imposed on the process

2b. Personal standards imposed on the process by the process operator(s)

2c. Psychological environment

5a. Process operator knowledge

5b. Operator's skills

5c. Operator's experienece

5d. Operator's physical attributes

4a. Facilities

4b. Equipment

4c. Tools

Customers Receive **Outputs**

1a. General description of the output

1b. Critical output requirements ("must haves")

1c. Important output requirements ("should haves")

1d. Other output requirements ("nice to haves")

Figure 3-8 Detailed Process Analysis Form.

Customers range from the next process through to the ultimate consumer of the process output, separately or as a component of a more complex product. Products come as a good, a service or a combination of both. A television is an example of a goods type product made up of many component parts. The delivery of the television to the ultimate consumer's home is a service type product. Painting a house is an example of goods (the paint and primer) and a service (applying the paint).

Tree Diagrams

Tree Diagrams expand a system or process in a logical fashion to expose all activities that make up the system or process. The example shown in Figure 3-9 will help explain a Tree Diagram. The system is a family, including their home. A family's critical processes make up the first expansion of the "limbs." Any one critical process can be expanded to see what key processes make up the critical processes. The key processes are then expanded to see process components or activities. The Tree Diagram can be taken to any level necessary to understand the system or process, or understand a critical component. Usually, the fourth or fifth levels highlight required activities.

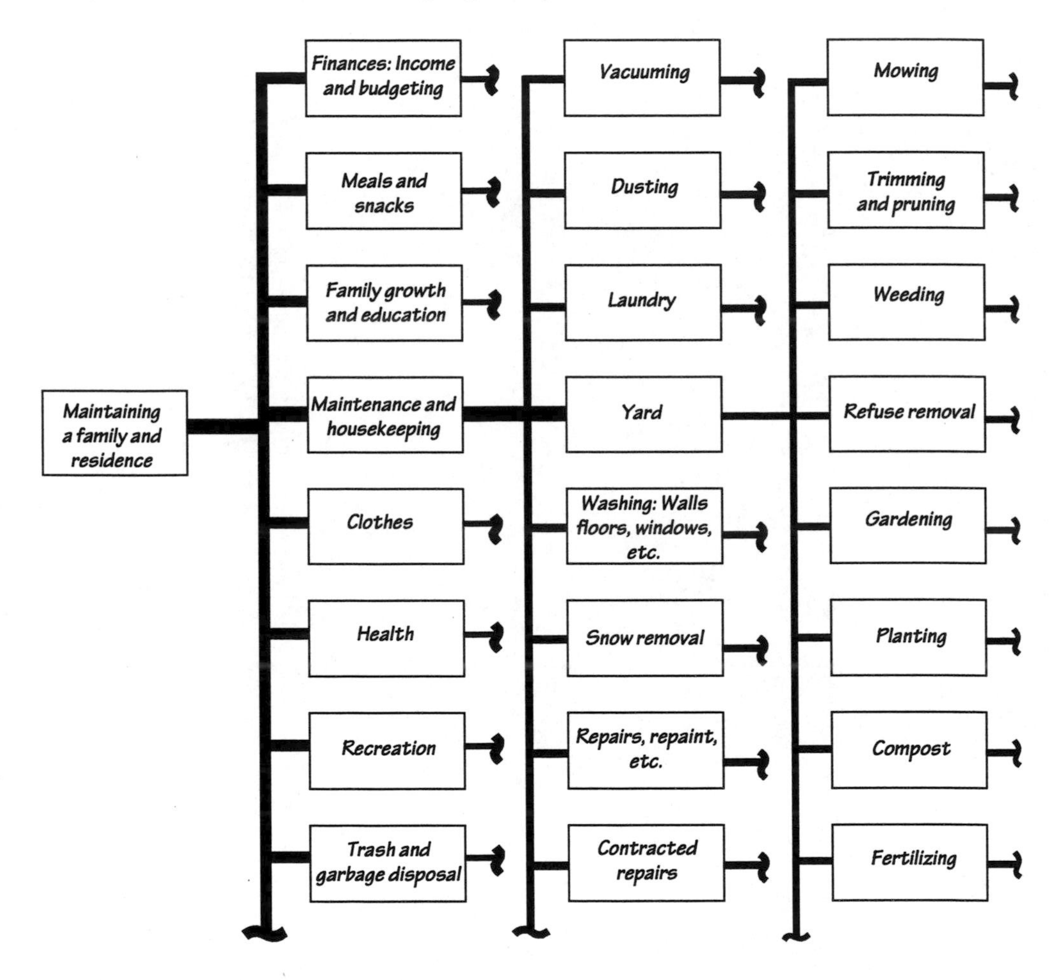

Figure 3-9 Tree Diagram of a family system.

Tree Diagrams become unwieldy at the third or greater levels. It is best to separately diagram each level with a large number of components. In Figure 3-9, the second level contains many components. "Maintenance and housekeeping" contains at least eight key processes. If the other seven critical processes contain eight, also, there would be about 64 key processes. For improvement purposes, a family might want to focus on improving a critical process such as home maintenance and housekeeping. The next level of the Tree Diagram notes the key processes making up "Maintenance and housekeeping." The key processes making up the next level of the tree shows areas of opportunity. If the yard needs improvement, the next level notes the processes making up "Yard." The family now focuses on the yard processes requiring improvement.

The Tree Diagram is a beneficial tool as individuals and teams set out to understand all the processes and activities making up a critical or key process. Post-it® notes are helpful and different colors could be used to note required or problem-prone processes and activities. If the team decides to Tree Diagram an entire system, it is then best to use a long wall and start at the top, branching downward. The advantage of diagraming an entire system is to gain a sense of everything which must go right to deliver the required output.

Chapter 4

Data Analysis and Presentation

Information Analysis

Data is not necessarily information. Data is a collection of numbers. Data requires conversion to information before most people can understand the meaning beneath the data.

Four convenient approaches are available for individuals and teams to use to turn data into information. All present the data in a graphical form. Histograms show how several categories of data or series of values relate to each other. Bar charts are the common forms of histograms, though lines often replace the bars, forming a line graph. Rearranging the bars of a graph according to descending values produces a Pareto analysis. The descending order of categories permits focusing on the problems with greatest potential. A run chart plots defects or variation over time. This permits viewing the output measure during a time period, providing a sense of the nature of the variation. A scatter diagram plots an independent variable versus a dependent variable. This provides a visual suggestion about how much the independent variable is affecting the dependent, and in which direction.

Histograms

Histograms present interval data in ascending or descending order in a bar graph form. Often, the Histogram's bars form a rough indication of a normal distribution. Output measures from other processes produce skewed distributions or distributions without a recognizable form. For more information on distributions, see the final topic in this chapter. The departure from a normal distribution also provides information on what the process being studied is doing or what has affected the process.

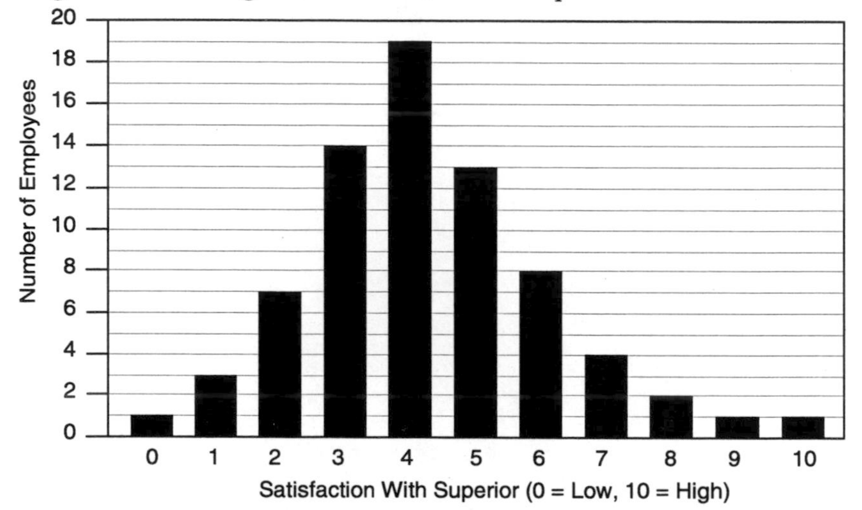

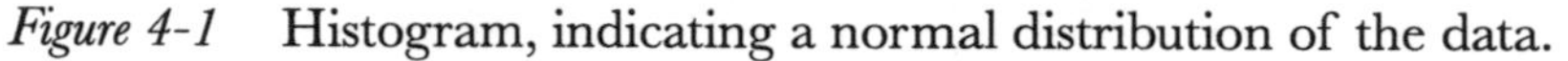

Figure 4-1 Histogram, indicating a normal distribution of the data.

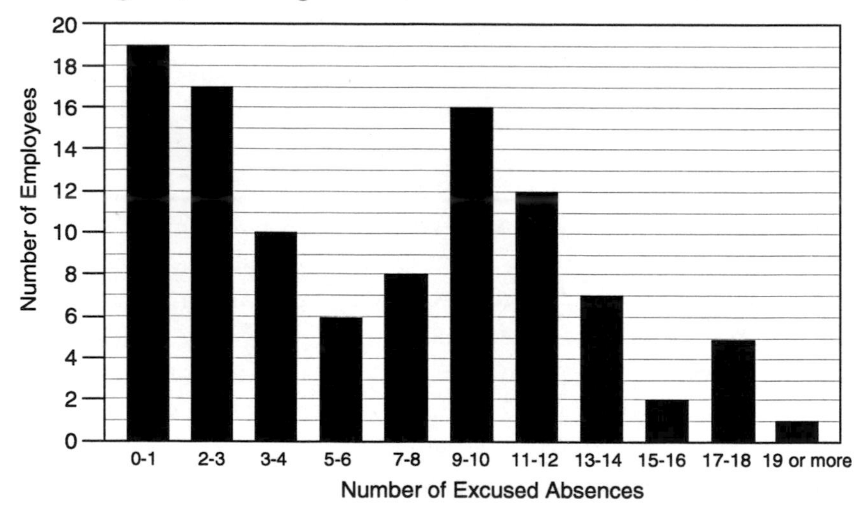

Figure 4-2 Histogram with each bar representing a range of values. Also, this Histogram does not form a relatively normal distribution as does the Histogram in Figure 4-1.

Pareto Charts/Analysis (80/20, Vital Few/Trivial Many, ABC Analysis)

An Italian economist and socialist, Vilfredo Pareto, discovered in the mid 1800's that 20 percent of the Italians held 80% of the wealth in Italy. More recently, statisticians and others have expanded his conclusion to a variety of related variables. For example, 20% of a company's sales force might make 80% of the sales. Twenty percent of the items an organization buys could account for 80% of the purchase expenditures. Eighty percent of the purchased supplies problems probably come from 20% the suppliers. It really is the 20% or the "vital few" which should command attention. The "trivial many" can consume an enormous amount of time with little pay back from improvements.

ABC analysis is an application of the Pareto principle. Since about 20% of the purchases account for 80% of the dollar value, concentration should be on the top 20%, the "A" purchases. The "B" items account for 80% of the remaining value (about 16%) and the "C" items are the remainder (about 4% of the value). The lack of a "C" item (a fastener, fax paper, a minor repair part or a latex glove) can shut down a process or slow down an organization. The lack of one of the "trivial many" may affect the goods or services provided to a critical customer.

The real value of a Pareto Analysis is the graphical presentation of the categories. Combine the categories in some logical manner to reduce the number to three to five. Three is common in purchasing or inventory analysis, though some companies use four. The first category should contain 70 to 90 percent of the cases. The second category should contain 70 to 90 percent of the remaining cases, and so on. The chart will look like the example shown in Figure 4-3.

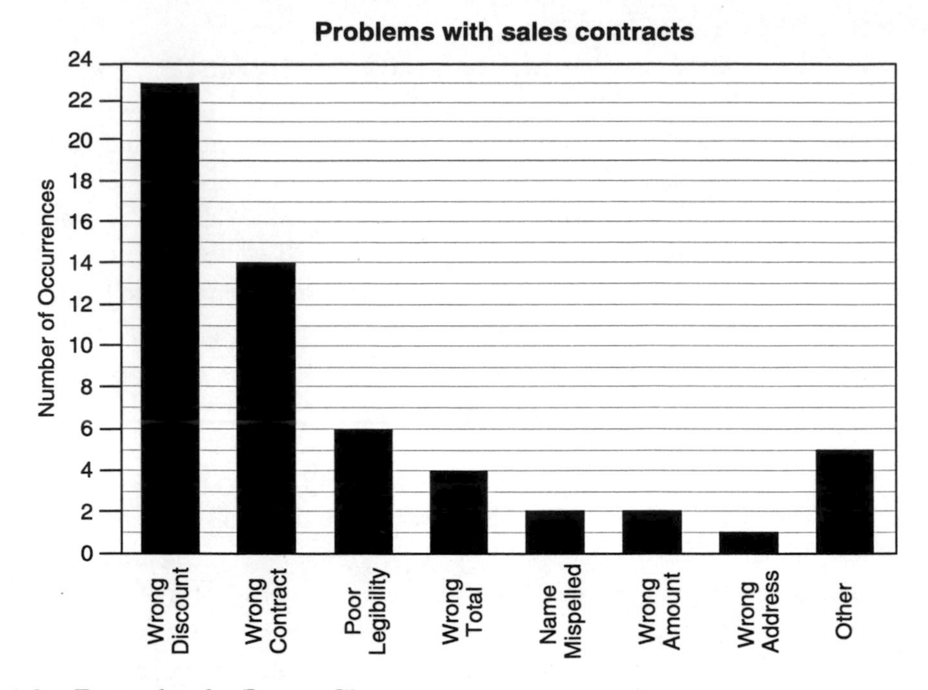

Figure 4-3 Example of a Pareto Chart.

Often, by grouping the defects to obtain one category with 70 to 85 percent of the defects, it becomes quite apparent where one should target the problem solving or improvement effort. A further Pareto Analysis of the largest category might uncover the one problem or defect category that is causing 80% of the 80% category (80% of 80% equals 64%). This would become the target for a problem-solving effort by a team or an expert. Pareto Analysis is an excellent tool for an individual or team to use in Step One of the systematic problem-solving approach presented in Section 2.

Run Charts

Run Charts plot an output quality measure or a process internal measure over time. The quality measure could be the number of defective outputs during a given period, the percentage of defects during the period or the deviation from a target value. The time measure could be any realistic period ranging from a minute to a year. After plotting twenty to thirty periods, the Run Chart presents a picture of the "health" of the process, as for the plotted quality measure.

The first decision to make is what quality measure to plot. One of the following three questions will point the direction:

1. What is most import to the customer? Convert the identified importance to a specific output measure.
2. Which measure, internal to the process, has the greatest effect on the output quality?
3. Which output measure causes the most concern, or predicts general quality the best?

Strive to include a measure from question one. A measure related to customer requirements provides the best information to plot. If this is not possible or practical, move to question two. Variation in this measure predicts variation in output quality. This measure is, therefore, one level removed from a customer satisfaction measure. Question three is the last resort because the measure does not directly relate to the output quality or the process customer.

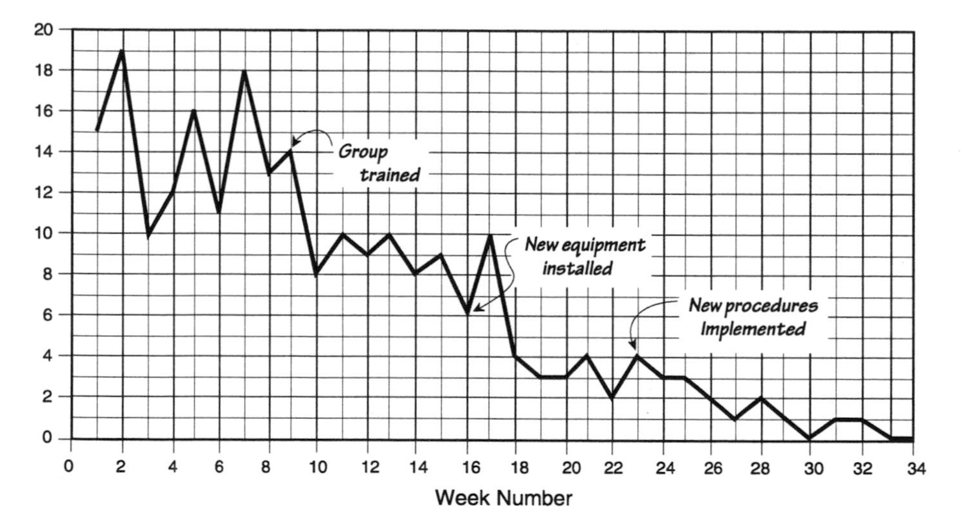

Figure 4-4 Run Chart indicating the results of several improvement efforts.

The presentation of the chosen quality measure depends upon whether the number of outputs during the period is constant. If the output is constant, plot the number of defectives. If the output varies, plot the percentage defective. If a degree of quality exists, as when the measure is the deviation from a target value, plot the actual deviation from the target value. Measures internal to the process can be any measurement: Temperature, speed, length, noise, etc.

The period chosen is a compromise between short enough periods to provide timely feedback and long enough to provide enough output quantity to represent the process. Plotting each minute would require almost constant measuring and would dictate automated sampling. A yearlong period would require a year between data points on the chart. Developing a sense of the process or changes in the process would take many years. Therefore, a time between samples must be chosen. A good rule-of-thumb is to obtain twenty to thirty outputs before plotting the process. The time scale would be a convenient period for twenty to thirty outputs to occur. Sometimes outputs occur regularly: hourly, daily or weekly. Examples of these would be sorting incoming mail, payroll activities, monthly accounting activities, serving lunch, etc. Plot each of these occurrences.

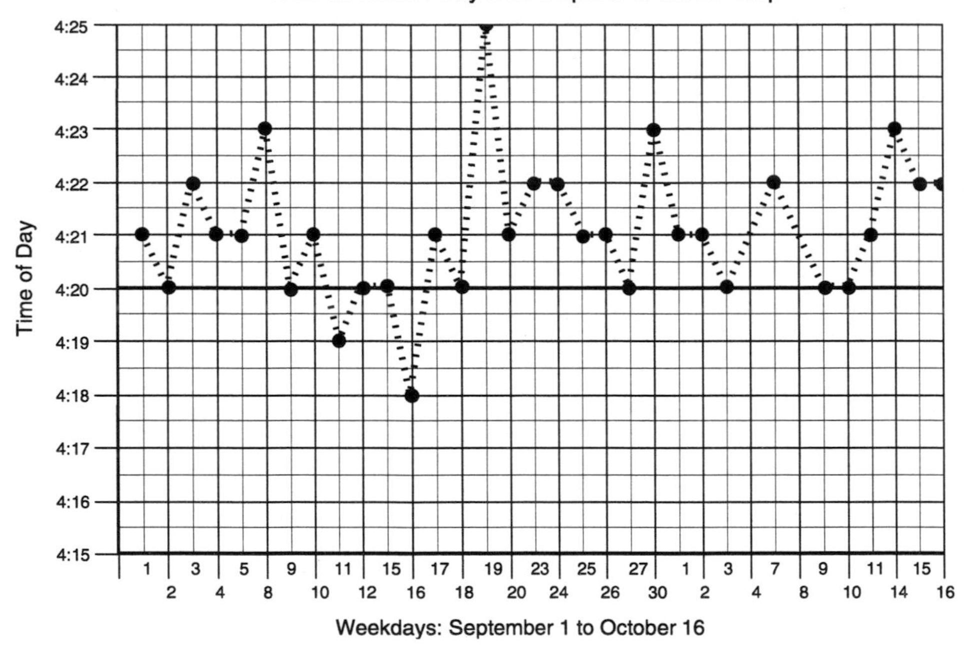

Figures 4-5 Example of a Run Chart using a measure other than defect number or defect rate

The Run Chart shown in Figure 4-5 plots the deviation from the scheduled (target) time a bus will depart at a given stop. The chart has been kept for thirty consecutive weekdays. The chart does not include the weekend days because the bus runs on a different schedule with dramatically different loads and traffic patterns. This would include measures from a different process. Do not combine two different processes. Therefore, plot weekends on a separate Run Chart. Chapter Five includes information on combined data and the requirement for stratifying combined processes.

The bus drivers collect the data, not to detect how well they are doing, but to determine the capability of the process, and to find special and common causes of variation. The chart also permits them to confirm that an improvement has improved a measure important to the customers, the riders.

Several focuses result from analyzing a Run Chart:

1. Is there an apparent cycle? In the chart shown in Figure 4.5, drawing in a dotted line between the Friday and Monday lines might be helpful. If a cycle is evident, what is the cycle produced by or related to? Business, yearly, seasonal, monthly, weekly and daily cycles are often evident in run charts. In the bus stop run chart, more people may leave early Fridays, producing a lighter passenger load that would permit the driver to follow the schedule better. Schedules are probably more difficult to maintain during winter in the northern climates.
2. Are there any unusual occurrences (special effects)? Is a point unusually high or low? What caused that incident? Are several (three or more) at the same value? Would one have expected this under normal circumstances? Is there an upward or downward trend? Is the trend desirable or undesirable? What could be causing this trend? Several factors could explain this variation. People get ill, take vacations and do move on. Replacements cause additional variation on the short term. Equipment breaks down and accidents do happen. In the bus stop Run Chart, an accident or equipment breakdown could have occurred, causing the bus to run late on September 19.
3. Is there a natural limit, beyond which measurements cannot occur? When filling a quart container, a natural limit exists on how much liquid the container can contain. The number of scratches on a new car delivered to a dealer cannot be less than zero. In the bus stop run chart, it is highly unlikely that any driver would depart much more than two or three minutes early.

4. Is there a legal, regulatory or customer-imposed limit to the measure? In the bus stop Run Chart, the bus must not leave the stop before the published departure time. Departure times after the scheduled departure are less critical to customers than departures before the scheduled time. This would cause drivers to cluster departure times on the late side of the published departure time. Interstate highways have speed limits. This would cluster measures at that speed or some speed a little higher.
5. Is there a relationship between the charted measure and another measure? Running some processes faster causes more variation. Changing lighting could cause process output changes. In the bus stop Run Chart, is there a relationship between considerably late departures and weather, passenger load or traffic.

Scatter Diagrams

Scatter Diagrams show how a dependent variable varies as an independent variable varies. The Scatter Diagram presents a picture of this variation and has two uses. First, it allows determination that the independent variable actually does affect the dependent variable, to what extent and in what direction.

Second, wanting a certain value of the dependent variable, a horizontal line can be drawn from that value to the central portion of the dots along that line. Drop a vertical line down to the horizontal axis. Figure 4-6 shows this method. The independent variable value at that point is the optimum value that will consistently produce the desired dependent value. The more the plot points produce a compact, straight line, the greater the relationship between the dependent value and the independent value.

Determining how baking time affects the crust on bread is an excellent example of how a Scatter Diagram is useful. Baking time could vary from just a few minutes to an hour, perhaps. Obviously, less baking time (the independent variable) produces a thinner crust (the dependent variable). Figure 4-6 plots an example of a Scatter Diagram for baking bread to produce a one-tenth inch crust, consistently. Note that the baker baked two different loaves at nine different times. The baker held all other variables constant. These variables included ingredient ratios, loaf size and temperature. The suggested baking time for a one-tenth-inch crust would be about 55 minutes.

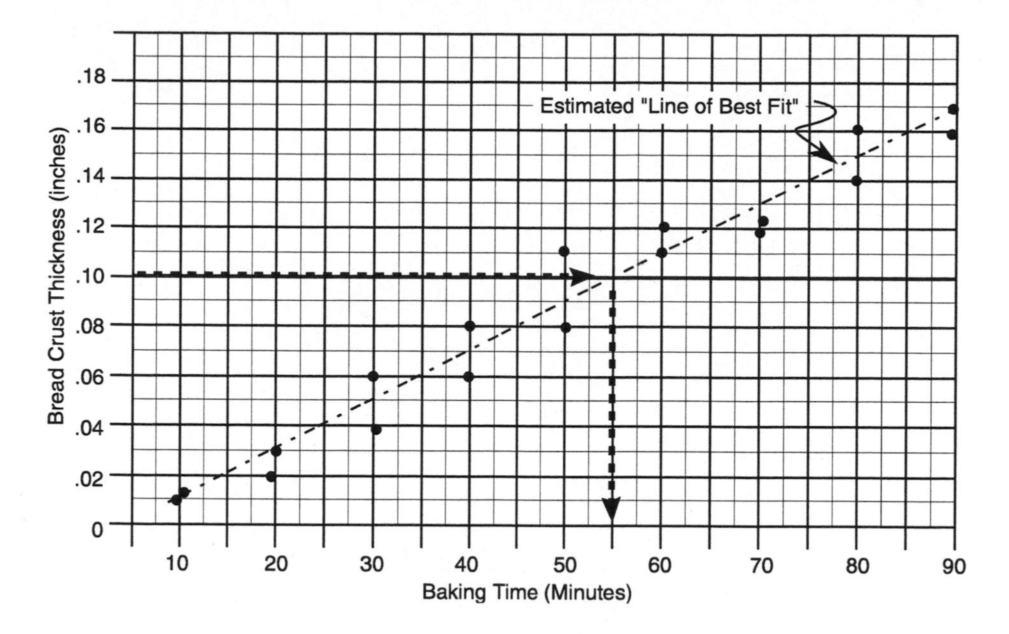

Figure 4-6 Scatter Diagram plotting bread crust thickness as baking time varies.

Figures 4-7, 4-8 and 4-9 each present a Scatter Diagram showing no relationship, a weak relationship and a strong relationship, respectively. The relationship in Figure 4-9 is a positive relationship. This means that as the independent variable increases the dependent variable increases.

Regression analysis, a statistical calculation, will establish the relationship between the independent and dependent variables. Regression analysis has two primary uses. First, the analysis will produce the formula for a regression line, or a "line of best fit." Refer to Figures 4-6 and 4-9 to see a regression line added to the scatter diagram. Second, the regression analysis will produce a statistical value that suggests the degree of statistical certainty to which a given independent value will produce a corresponding dependent value. Use the formula for the "line of best fit" to calculate the independent value that produces a desired dependent value. Regression analysis is beyond the scope of this book. Most statistical books contain information on regression analysis. However, a line can be placed within a number of dots, visually, to form an estimated line of best best fit.

If the plotted values form a pattern resembling a straight line, as in Figure 4-9, a line can be drawn to estimate the "line of best fit." It is critically important to recognize that the line drawn without the benefit of regression analysis is an estimate, only. It is, however, close enough to gain a sense of how the process is operating.

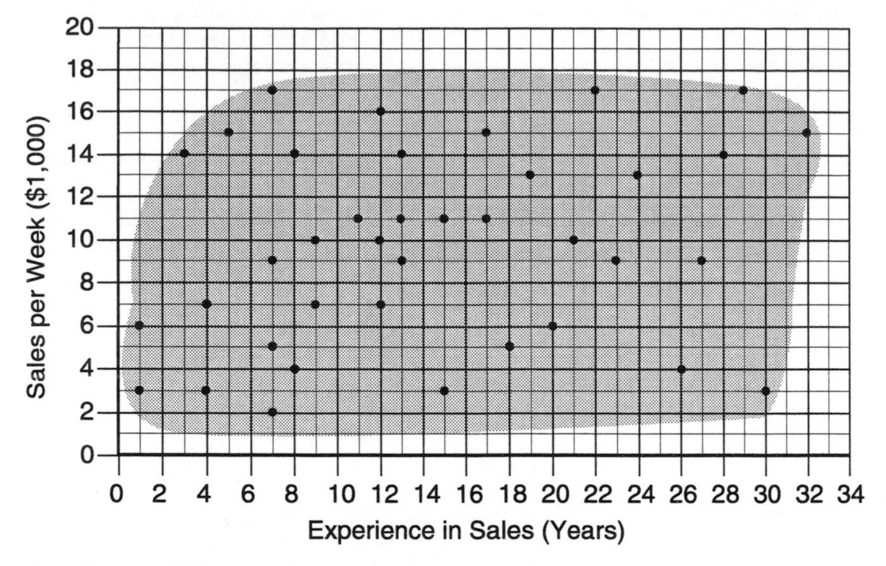

Figure 4-7 Scatter Diagram indicating no relationship between the independent variable and the dependent variable.

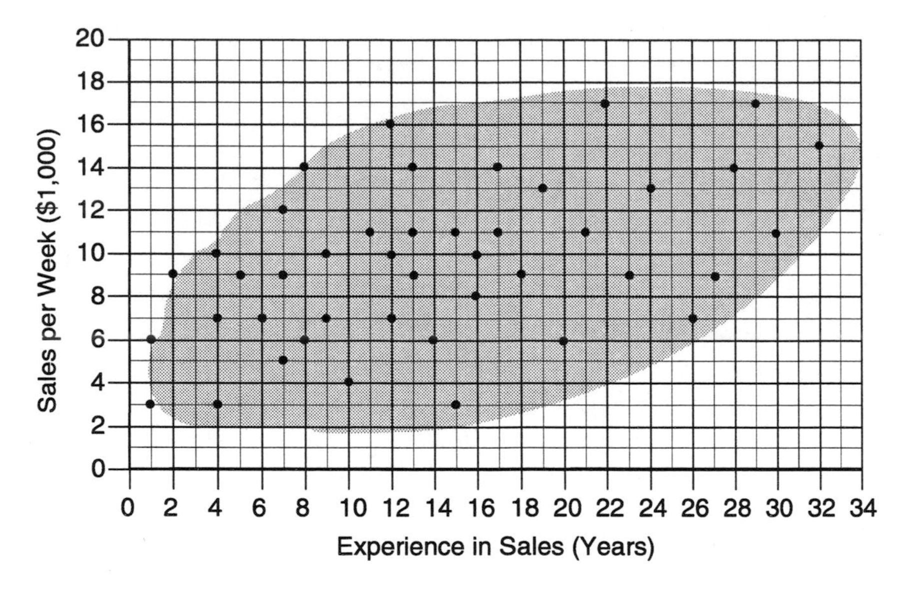

Figure 4-8 Scatter Diagram indicating a weak relationship between the independent variable and dependent variable.

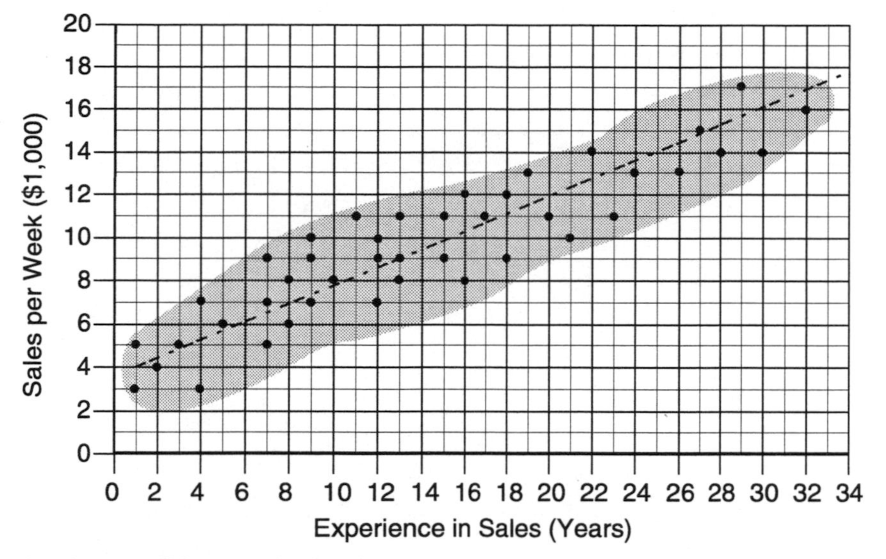

Figure 4-9 Scatter Diagram indicating a strong relationship between the independent variable and the dependent variable. A "line of best fit" has been drawn in to suggest the relationship.

Useful Statistical Thoughts

A variety of statistical approaches are useful as teams and individuals set out to eliminate problems and improve processes. Chapter One touches on the critical need to sample randomly, a cornerstone for gathering data and analyzing the data statistically. Chapter Two condenses statistical approaches to simplify charting, formulas and decision-making activities. The concept of standard deviation (sigma) is also presented in Chapter Two.

Other concepts from the field of statistics are also helpful. Most of these are useful in daily work life as one strives to understand how a process is behaving based on sample data. One could think of these useful statistical thoughts as mechanisms for allowing the data to speak in terms humans can better understand.

Populations and Samples

The population is the total output of a process over a period, all employees of a company, a complete shipment of supplies received, all sales or customers during a given period. Sometimes one need not sample the population if the population is small. Other populations may be so large, or the analysis of each individual so costly or complex, that analyzing part of the population is the only alternative. The portion analyzed is the sample. Gather samples at random to make certain that the sample group represents the population as much as possible.

How many samples should one take to be certain that the sample is a reasonable representation of the population? Complex statistical approaches exist that help determine what size sample to take from a given population size to be certain of the results at a given level of confidence. For problem-solving and process improvement purposes, some simple "rules-of-thumb" are useful. As few as twenty samples provide a general, though minimal, knowledge of the population. Thirty samples improve the confidence of the analysis, allowing non-critical decisions based on the data. A sample of fifty improves confidence to the point that the sample reasonably represents the population. If a high level of confidence is necessary, take a sample of one to two hundred.

Samples of One

Though it should be obvious, a very small sample, especially samples of one, provides poor problem solving or process improvement information. Decisions based on samples of one are too common, and may do more harm than good. A sample of one, or even a few samples, from a population of any size cannot possibly represent the population. One serious customer complaint does not mean that all the goods or services of the type delivered to that customer are defective. What one customer complaint does suggest is that one customer has a complaint. One extreme disappointment with a product often leads a person to translate that experience to all of that product. This explains one reason for the existence of prejudices against certain type people, companies, products, methods, and even improvement approaches. "We tried that years ago and it did not work then so it will not work now!" is a conclusion based on a sample of one. A sample of one often explains why special causes (one unique occurrence of a problem) should not be corrected as a common cause. If the problem has occurred once, only, it may never happen again. Corrective actions on the system would add cost and could even increase variation.

Average (Mean and Median)

Average is one of the most commonly used statistical terms. The number we usually associate with the word "average" is most often the mean of the set of numbers. Compute the mean by first adding all the values of the data set. Divide this number by the number of values in the data set. For example, the mean of the data set 5, 4, 6, 5 and 8 is the sum of these five numbers divided by five. Twenty-eight (5+4+5+6+8=28) divided by five equals 5.6.

"Average" may also refer to the median or middle number in the data set, after arranging the numbers in descending or ascending order. The above set of five numbers arranged in descending order produces this set: 8, 6, 5, 5, 4. Five is the middle number. The median for this data set is five. If a data set contains an even number of values, then the median is half way between the middle two numbers. For example, the median of 4, 5, 5, 7, 8, and 9 is 6, the number half way between five and seven. For more complex numbers such as 4.5, 5.7, 5.9, 6.3, 7.3 and 8.0, add the two middle numbers and divide by two. Therefore, 5.9 plus 6.3 equals 12.2, and 12.2 divided by 2 equals 6.1. The median of the data set is 6.1.

One must be extremely careful not to put too much importance on an average. Always remember the saying, "Put your head in the furnace and your feet in the freezer, and on the average, you will be comfortable." Another way of looking at the problem is that a median tells one that half the numbers are larger than the median and half are smaller. The median does not suggest the magnitude of the other numbers. The median does not show the distribution of the numbers. Given the numbers 5, 6 and 7, the median is six.

Surprisingly, if the numbers are 5, 6 and 100, the median is still six. A similar problem exists with the mean. The mean indicates an arithmetic average, not a position average as the median does. Given the numbers 5, 6, and 7, the mean is six. However, if the numbers are 5, 6 and 100, the mean is 37.

Central Tendency

Average is useful in understanding the central tendency of a data set. How a process varies on either side of a central value is critical in understanding how the process is behaving. Mean is the best average if the data forms a normal or symmetrical distribution. See Frequency Distributions, later in this chapter. Median is a better measure of central tendency if the data forms an abnormal distribution.

Numbers can misbehave, though. Consider a workgroup made up of five employees with hourly wages of $6.50, $6.75, $7.10, $7.75 and $12.50. The mean wage for the group is $8.12. Eighty percent of the workgroup (four of five employees) would wonder how the mean wage could be higher than most of the workgroup. Even the median wage of $7.10 does not really represent the average wage of this group. Now, assume that the employee making $6.50 per hour receives a raise to $6.75 per hour. This does not affect the mean as much as it does the mode. Two wages are now $6.75 per hour, the typical wage of this workgroup. Three members of the workgroup would wonder how the wages of two employees could represent the typical wages of the entire group.

Fortunately, larger data sets greatly reduce the effect of one or two abnormal values. This is why one should accumulate 20 to 30 samples before performing any statistical analysis of the data. Of course, this might not be possible if the total population is fewer than twenty. Always be on the alert for abnormal values in small data sets and the profound effect they may have on the analysis.

Mode

Mode is simply the most frequently occurring value in a data set. In the data set 4, 5, 6, 7, 7, 7, and 8 is the mode is seven. If two different values occur most frequently, the data set is bimodal. Mode must never be confused with an average value. Mode merely indicates the most frequent value.

Range

Range is the spread of the data set. Compute range by subtracting the smallest value in the data set from the largest. The range value has several uses. The formulas for the upper and lower control limits (Statistical Process Control, Chapter Two) use the range. If the distribution is a normal distribution from a stable process, divide the range by six to arrive at an estimated standard deviation. This is because 99.74% of the data values will fall between minus three sigma and plus three sigma. Only .26% of the time will a one find a value less than three sigmas below the mean or more than plus three sigmas above the mean, or a six-sigma spread.

Frequency Distributions

The frequency distributions of data sets can form a variety of distributions. The distribuions can be normal, skewed or abnormal. Figure 4-10 compares different type distributions.

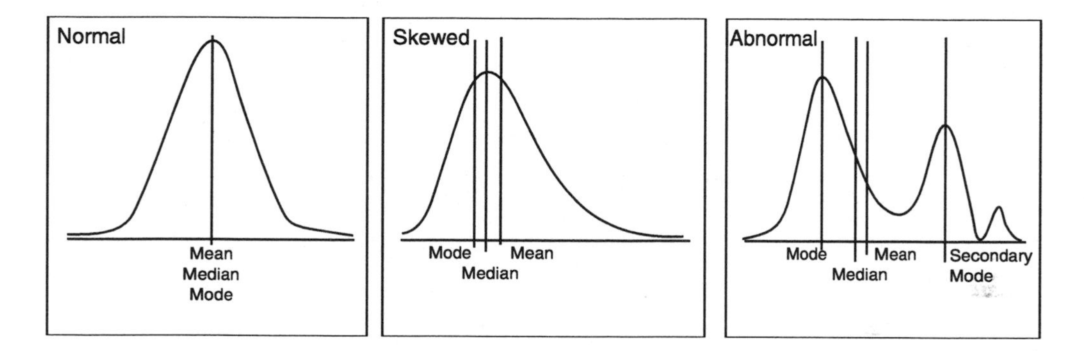

Figure 4-10 Normal, Skewed and Abnormal distribution.

Normal distributions are the distributions of choice as teams and individuals engage in process improvement. Normal distributions generally have an absence of special causes, the variation resulting from random variation, only. Normal distributions produce much more predictable outputs. Normal distributions have coinciding means, medians and modes.

Variability

The measure of variability is the standard deviation of the data set or sigma (σ). The easiest way to imagine sigma is to consider the value as representing how far one can go on either side of the mean to capture about two-thirds of the data in the data set, in a normal distribution. In other words, if the mean is 100 and the sigma is 10, the range of values from 90 (100-10) to 110 (100+10) would include about two-thirds (68%) of the data. Two sigma above and below the mean includes about 95.5% of the data. Three sigma below the mean to three sigma above the mean includes almost 100 percent of the data. (The actual figure is 99.74%.) The area between minus four sigma and plus four sigma contains 99.994% of the data.

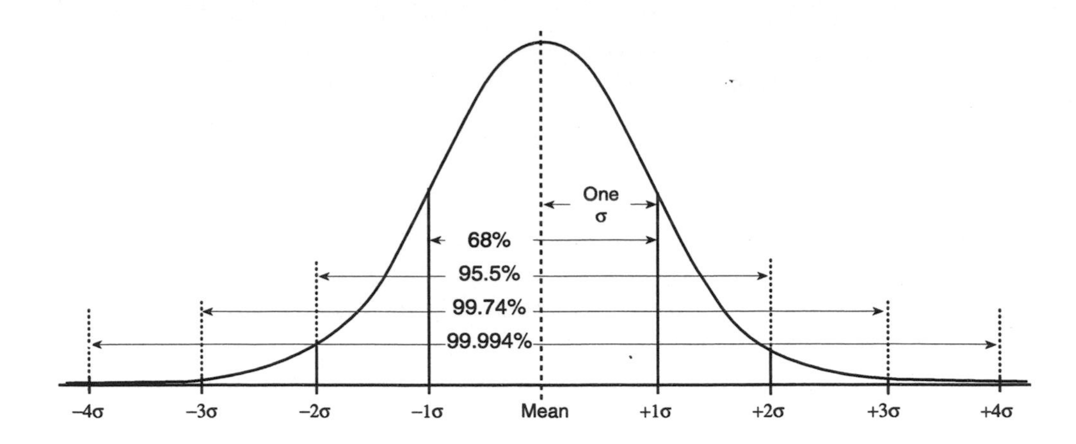

Figure 4-11 Proportion of data included in a given sigma range.

In fact, knowing the mean and sigma, then estimating the range is easy. If the mean is 100 and sigma is 10, the data extends from about three sigma below the mean (70) to three sigma above the mean (130). Range, in this example, is about 60. This approach can also be reversed. If the range is known and the data forms a normal distribution, compute sigma by dividing the range by six. Example: The data forms a normal distribution, the lowest value is 210 and the highest value is 255. The range is 45 (255-210), so the estimated sigma of this data is 7.5 (45 divided by 6). If the distribution is not normal, this approach does not work.

Probability

Probability forms a foundation of statistics and is generally expressed as a ratio. Assume that six different outcomes are possible and each has an equal chance of occurring. Then, any one outcome will occur about one time out of six or one-sixth of the time. A deck of cards has 52 different cards. Pulling one card at random and getting a five of diamonds will happen once out of every 52 times. The chances of getting a five of diamonds is one in 52, one of 52, 1/52, 1:52, or 1.9% of the time. Another example is flipping a fair coin. The outcome is "heads" or "tails." Half the time (1 out of 2 or 1/2) the coin will come to rest "heads up." Flipping the coin twice and expecting "heads" on both will be 1/2 of 1/2, or 1/4. Two "heads" in a row will occur one out of four times.

Probability allows for some unusual quirks. Since each toss of a fair coin is a new event, the coin can come up a head or a tail. Just because the first toss produces a head has nothing to do with what the second toss will produce. However, over many tosses, the number of tails will roughly equal the number of heads.

In the everyday business world, little is known about any particular outcome to predict, with a high degree of certainty, what percentage of time a certain desired or undesired outcome will occur. However, stable systems do allow for educated guesses. The past relative frequency of occurrence does allow for the prediction of future events. The problem is that few systems or process are stable over very long periods. Often, data is lacking, the system or process is new, or conditions are unstable. The manager or process operation must then resort to subjective probability.

Three types of probability exist. Classical probability allows for accurate knowledge of the probability of an outcome. Relative frequency of occurrence probability provides for estimates based on past data. Subjective probability is a guess based on very little data. Though some incidences of the application of classical probability in improvement efforts occur, relative frequency of occurrence is the probability most applicable to problem-solving and process improvement.

Classical Probability

Classical Probability deals with known proportions. This is because a careful analysis of the results of many trials or occurrences has taken place. A coin, die, deck of cards, specific shipment of finished goods or the next 200 customers to buy a specific product are examples of possibilities that could be suitable for classical probability. Assume that a company receives an order of 100 pens, one of which is defective. The chance of getting the defective pen by drawing 10 pens at random is 10 in 100 or 10% or 1/10 probability. In other words, a 90% probability exists that all 10 pens in the sample will be good.

Relative Frequency of Occurrence

Relative Frequency of Occurrence probability is based on the fact that with many occurrences, one can predict the likelihood of a given outcome, if conditions are stable and will remain stable. Actuarial tables are based on relative frequency of occurrence. For example, total the number of men who died while age sixty during the past 10 years. Divide this number by the total number of men who were age sixty during the past ten years. The result is the probability of a sixty-year-old man dying while age sixty. This assumes a stable system for sixty-year-old men. That is, nothing has changed to increase or decrease the death rates of sixty-year-old men.

Subjective Probability

Subjective probability is based on personal or group beliefs using whatever evidence is available. Lacking conditions or data to use classical or frequency of occurrence probability, subjective probability is the only alternative. For example, a sales manager has narrowed all the applicants for a sales position to three prime candidates. What is the probability that each will relate positively to the company's customers? Precious little data exists to support the probability assigned to each. Collecting data to support relative frequency of occurrence would take years, and many hired and fired salespeople. The probability assigned to each is highly subjective based on some objective data. Subjective probability will predict the probability of a new product selling well, without supporting data.

What Sample Size to Take?

Often, the question arises concerning a minimum sample size. For most problem-solving and improvement efforts, a sample size of 20 or more from a population of several hundred will be sufficient to make an estimate about the population. A sample size of at least 30 will produce a fair comfort level that the sample represents the population. A sample size of 50 increases comfort to an even higher level. If the sampler takes two or more samples from each period of a process, sample size is for the number of multiple sample lots, not the total number of samples taken.

However, for larger populations and cases where given level of confidence about the sample size is important, one must resort to alternative approaches. Three approaches allow the individual or team to figure out what sample size to take:

1. If conditions are stable (predictable, no special causes of variation), and the sampler knows (with relative confidence or can estimate fairly close) the standard deviation of the population, a formula will help:

   $n = (ZS/l)^2$ Where: n = Sample size (Minimum)

   Z = 1.96 for 95% confidence

   2.58 for 99% confidence

   S = Standard Deviation of population

   l = The maximum amount of desired or allowable deviation between the sample mean from the population mean. For example, if the actual mean age of your population of customers is 37, can you tolerate a sample mean that is as much as 3 more or 3 less? If so, $l = 3$.

   Note that population size is not considered using this approach.

2. If certainty exists that the population parameters of the process under investigation are stable, but you do not know the standard deviation of the population, take a sample size using these criteria:
   a. Take the square root of the population being studied and round up to the nearest 20:

      $N = 200 \quad \sqrt{n} = 14$, take a sample of 20

      $N = 1000 \quad \sqrt{n} = 32$, take a sample of 40

   b. Compute $\overline{X}_1$, R_1, and σ_1 for the first sample. Return the samples, if using physical sampling, and take a second sample of the same size. Next, compute $\overline{X}_2$, R_2 and σ_2 for the second sample. How much do the samples differ? What difference can you tolerate between the samples for adequate problem-solving and improvement decisions?
   c. Then:
      - If the conclusion is that the samples are close enough, add them and compute a new means, range and standard deviation of the sum of the two samples.
      - Or, if the conclusion is that the two samples are significantly different, and/or there is not comfort with the results and sampling is quick and inexpensive, then, increase the sample size.
3. If one does not know the standard deviation of the population and that the conditions are stable, or if absolute certainty about the required sample size is a necessity, now is the time to contact a statistician for assistance. Determining the sample size to take will be complex.

Chapter 5

Making Sense of Data: Changing Data to Information

Again, Data Is Not the Same as Information

Often, data is no more than a collection of numbers, words or thoughts. This data may represent a situation or what has happened, but may not be in the necessary form for an individual or team to make informed decisions about correction or improvement actions.

This chapter offers four non-statistical approaches to help individuals and teams in making sense of data. Affinity Diagraming reduces large amounts of verbal data to manageable groups with some common element. The Relational Digraph may sound a bit forbidding, but is easy to use and does help in understanding the relationships between clustered information or between key issues. Use Stratification to separate data into more meaningful subgroups. Matrix Diagrams and Synthesis Matrices improve understanding by relating two sets of data. Matrix Diagrams also relates a set of data with a set of standards, issues or other important criteria.

Affinity Diagrams: Clustering Verbal Data

Affinity Diagraming is a close derivation of the KJ Method developed by Dr. Kawakita Jiro, a Japanese anthropologist. The technique is an excellent approach for a group to use to sift through large quantities of verbal information or verbal data to identify patterns, similarities or relationships. This reduces the collection of more diverse data to a few clusters of data. Then, give each cluster a title or header describing the relationship among the data within the cluster. Affinity Diagraming is loosely considered as one of the *seven management tools for quality improvement.*

The Affinity Diagram approach begins with a large quantity of verbal information, recorded in meaningful phrases, and written on note cards or Post-it® notes. Placing the Post-it® notes on a wall allows the entire team to see all the notes and move them about with ease. Note cards permit clustering on a table, with a header card placed on top of each stack of note cards.

The three primary uses for the Affinity Diagrams are:

1. Organizing ideas and information to make better sense of an issue,
2. Assisting in defining the nature of the problem, and
3. Pointing the team or individual in the right direction for problem-solving, process improvement or creation of a new process or product.

The following eight-step approach might suggest that clustering is a complex tool. However, the first two steps merely generate the ideas and document them on note cards or Post-it® notes. These first two steps tend produce better results using a team, if the verbal data has not already been collected. The third step is a mechanical step, recording the statements on individual cards. Step Four is the actual clustering step and Step Five creates headings or titles for each cluster of data. Step Six recommends a second pass if more than five or six clusters remain. The seventh step cleans up the data within each cluster, organizes the clusters and records the results. Step Eight repeats the process to see if another set of clusters is possible.

Step One: Select a subject, theme or issue. This statement:

- Should be clear to all team members.
- Represents the direction the group wishes to take.
- Should be recorded on the top of a flip chart as a constant reminder of the theme.

Step Two: Collect the verbal data:

- Generate or collect phrases.
- This "raw" data represents facts, ideas, inferences, predictions, but not necessarily solutions or goals.
- The team could use a brainstorming approach to generate the data for the data cards; or the data could come from individual observations, comments on customer comment cards, focus groups or a variety of other data collection approaches.
- Individuals using Affinity Diagraming may have the verbal data and want to understand the meaning of the data through Affinity Diagraming. Affinity Diagraming loses its power if only one person is involved. Therefore, involving others in the clustering exercise is highly beneficial for an individual.

Step Three: Make the data cards or Post-it® notes:

- This step could occur during the generation of the data.
- Each card or note should capture the essence of the raw data, not extensive details.
- All team members must agree that everyone understands the intent of the phrase.
- Keep personalities out of the data cards at all cost.

Step Four: Arrange the data cards:

- Spread the cards out on a table or line up the Post-it® notes on the wall in a single line. The arrangement must be in random order at this point.
- The team leader should read each card to the group. This helps eliminate the possibility that some team members may skip some cards. Also, reading each card allows team members to ask for clarification.
- The team leader asks members to think about which cards have similar meaning or a relationship to another card. Which cards say the same thing in different ways? Is one a subset of another? Always attain a team consensus any time team members combine two or more cards.
- Group identical or similar cards to make the data more manageable.
- The team members collectively, and without speaking, group cards with a relationship or similarity. This can be done individually between meetings or during the team meeting as a team activity. Use one of the following approaches:

- The entire team rearranges the cards, collectively. (This is appropriate for smaller groups.)
- Subgroups, one subgroup at a time, cluster the cards, without conversation. (This works better for larger groups.)
- Team members, individually, take turns clustering the cards until all members pass. (Good for smaller groups.)
- The leader can ask the team for suggested groupings. (Works with any size group, but is much less effective in obtaining breakthrough clusters.)

– Do not force single cards into a group unless a definite relationship exists. Let the single cards remain as "loners."

– To handle individual disagreements, the team leader could say:
  - "If you don't like where the card is located, move it!"
  - "Why don't the two of you work it out, off line?"

  However, do not waste too much time discussing or arguing about right or wrong clusters at this point.

– Occasionally a team consensus that the Affinity Diagraming exercise is not complete at the end of the meeting does occur. The cards could remain in the meeting room or another common place for a short period for team members to continue the clustering process, or the exercise could be completed at the next meeting.

– Repeat the process until the team forms five to ten groups.

Step Five: The team makes "Header Cards" to suggest relationships and then collects the clustered data cards underneath the appropriate header:

– Header cards title the groups.

– The statement must be clear and concise, and yet capture the essence or common thread of the data cards in the group. The header card:
  - Should not contain "fuzzy" language, jargon, or cliches. Avoid stereotypes.
  - Reflects the nature of the common thread or common bond between the data cards.

– The header cards often represent breakthrough thinking.

Step Six: Second pass, if the problem or issue is critical:

– Repeat steps four and five to see if the initial relationships are appropriate and to identify the relationships within a group: subgroup relationships.

– It would be optimum to have five or fewer major groups, but this does not always happen.

– Consider what would happen if the team tightens or loosens the operational definition of "relationship," "similarity," "common" or "pattern."

Step Seven: Neaten up the clusters and record the results of the Affinity Diagraming session:

Neaten:

– Arrange the data cards underneath the header card, trying to develop a logical order of data cards under each header card.
– Put the clusters in a more logical order.
– Tighten up the vertical spaces and straighten the horizontal lines.
– Make certain enough space is present between the groups so that each cluster is obvious.

Record:

– Capture the theme, data and relationship statements exactly as on the cards.
– Draw a box around each group, including the header card.
– A variety of computer programs could help the recorder.
– Reproduce and send to team members.

Step Eight: Discover other clustering possibilities:

– This is an optional, but recommended step, particularly if the team is studying a critical issue or problem, or if a team member has concern that the team has not arrived at a team consensus.
– Return to Step Six and randomize the data cards.
– Repeat Step Five.
– If a new clustering scheme emerges, continue through Steps Six to Seven.

Figure 5-1 shows the results of a team using the Affinity Diagraming tool to cluster the comments written in on the benefits section of an employee survey. Note that the common thread used was the type of benefit. Another common thread that could have emerged is type of comment: "Not enough information available," "Lack of training," "Age or seniority related," and "Administrative problems."

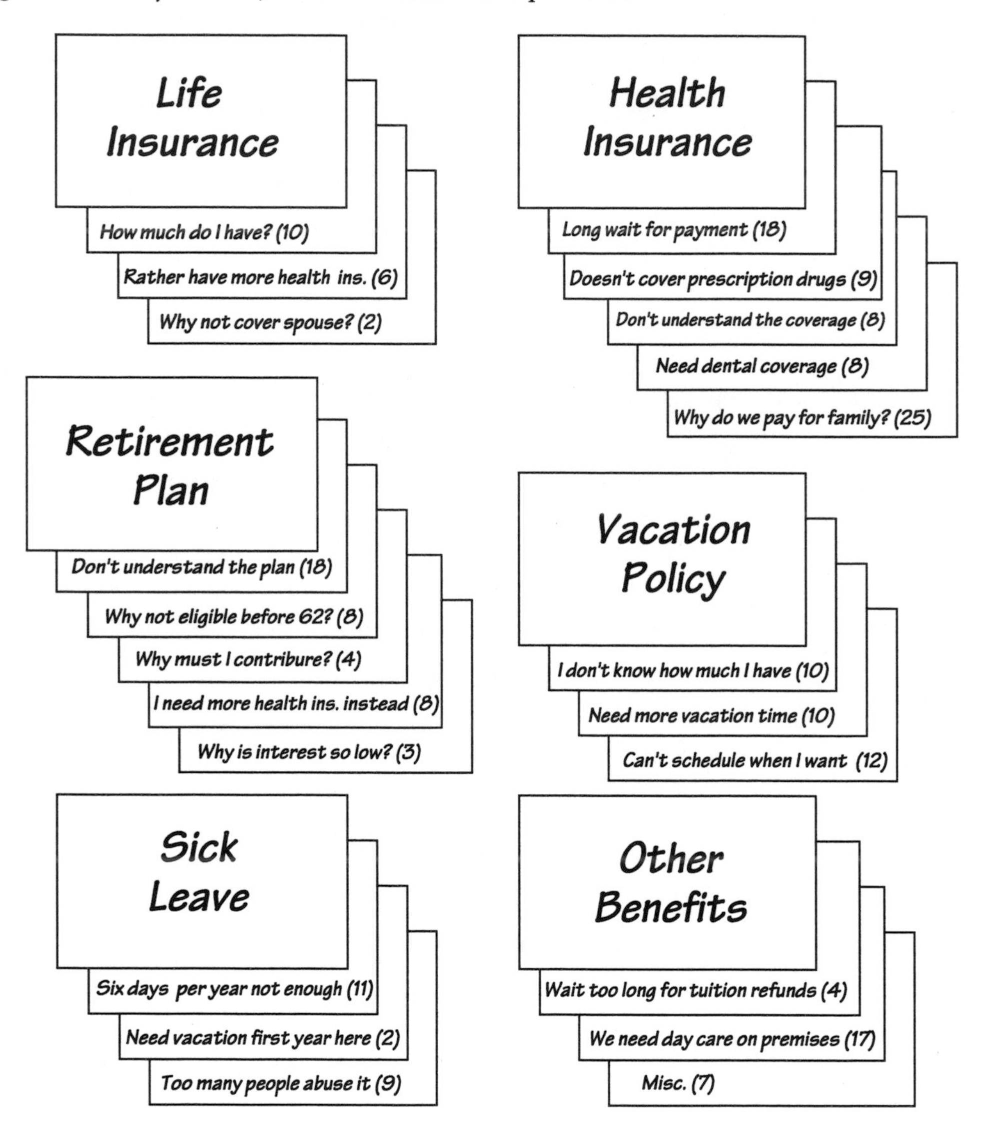

Figure 5-1 Affinity Diagraming (Clustering) the comments in the benefits portion of an employee survey.

Relational Digraphs

Sometimes the result of a team Brainstorming and Affinity Diagraming effort will be a group of header cards. This could also occur when an individual considers several issues. At this point the Relational Digraph is a useful tool.

The Relational Digraph is a simple technique for visually presenting the relationships among the groups or issues. If one group or issue causes or precedes another, a line is drawn to it with the arrowhead pointing to the caused or later one. When the identification of all relationships has been completed, the Relational Digraph will appear similar to the one in Figure 5-2.

The groups or issues could relate to each other in a variety of ways. Deciding the type relationship between the headers is critically important. The possibilities are:

- Time Sequence Which occurs before the other? Does the proper completion of one make a difference to the completion of the next? An example might be that to interview job applicants, the application must be completed. To get applicants in to fill out applications, one must announce the position. One cannot announce the opening until a requisition is completed and approved.
- Cause and Effect This is similar to the "Time Sequence" but clear cause and effect relationships are not completely dependent on time. For example, a poorly completed job opening requisition will cause the announcement of the wrong job and the wrong people to apply.
- Influences An alternative to the "cause and effect" relationship could be an influence or impact relationship among issues, concerns, problems or clustered groups. Simply ask of each issue: "What other issues are influenced or affected by this issue?"
- Information Flow Showing relationships based on information flow is not often used, but could provide a meaningful way to understand the grouped data. Often, the identification of information bottlenecks or distortion points is possible.
- Work Flow A Relational Digraph showing work flow is a simple one. Sometimes the work flow approach points out bottlenecks and problem areas. Adding information flows is often beneficial. Connect header cards representing key work processes by one type of line (solid) representing the flow of work and another type line (dotted) to show information flow. This will highlight any areas lacking a necessary information flow in connected processes or activities.

- Driving Force Sometimes teams find that identifying driving forces among the groups of data or issues is interesting. Draw a line from one group to the others it drives, with the arrowhead pointing to the driven one. The driving force group or issue is the candidate for attention. An example of a driving force issue is pay. Pay translates into status, a sense of worth or contribution, and many other personal aspects of a person's life. High needs for status may be the driving force for more pay, and a high need for pay may drive actions a person takes at work.

When the Relational Digraph is completed, it is time to identify the focus of a problem-solving or improvement project. The group or issue with all lines pointing away from it is the most likely candidate. The groups or issues with only one line pointing to it may be a candidate for a quick fix or a temporary correction. In Figure 5-2, the group of complaints labeled "Company seems too worried about controlling costs" points to five other groups. The team should address this issue. The collection of issues and complaints headed "Customer never taken seriously" points to five others and had one pointing to it. Perhaps something should be done to improve this situation, though the team should consider the effect of eliminating "Company seems too worried about controlling costs" may help a great deal.

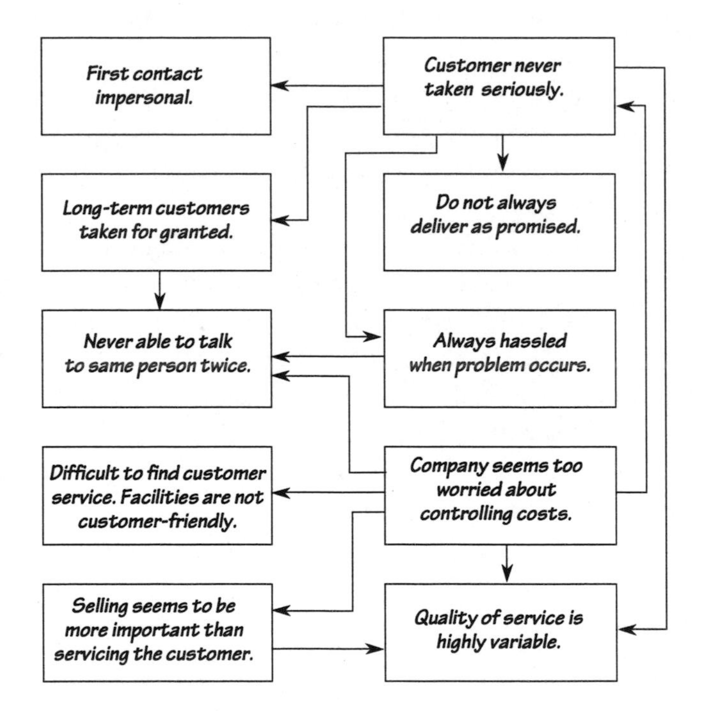

Figure 5-2 Relations Diagraph of clustered customer opinion data about customer service

Stratifying Data

Often a data set consists of two or more subsets of related data. When this is the situation, stratifying the data is extremely beneficial. Stratification separates the data into two or more homogenous sets of data. The resulting sets of data can then be analyzed using a variety of approaches, the most common being the run chart or frequency distribution. Compare different strata to learn what is affecting each set and what affects the master set.

In stratifying data, a team or individual could use many different criteria. Some possibilities are:

- Types: Try different types of customer complaints, defects, or problems. Strive to keep the strata down to two or three, and no more than four. Group types if necessary. For example, stratify the defects in newly delivered vehicles using visual defects as one strata and mechanical defects as the other. Another stratification approach is to separate automobiles and pick-up trucks.
- Location: Stratify physical or functional areas. Areas of the world, nation, region or state are possibilities. Locations of the organization, where sales occur or where use is made of the product are other ways to stratify. Material from two different sources could be a reason to stratify. One could stratify the physical locations of defects on the product. For example, this would provide the necessary understanding of how the data behaves relating to damage on the top of cartons as compared to damage on the sides.
- Time: Time often makes a difference in the data. The shift, day, week, month or season during which the product was produced or delivered could be a possibility. Using "shift" as an example, is there a difference in the data when separated by shift? Also, consider lengths of time or time delay issues. Is customer satisfaction different 90 days after delivery compared satisfaction at time of delivery? Stratification answers these questions and provides a basis for comparison.
- People: We have known for a long time that people make a difference. Nevertheless, could the data show this? Stratifying data based on whom or what group is associated with the data often provides insight into people differences. Do seasoned sales people produce different results, consistently, than novice sales people? Does a college degree make a difference? Stratifying the data relating to people with a college degree and people without could answer this question in relationship to a certain job or activity.

- Equipment or Tools: Different tools or equipment types used to produce or deliver the product could make a difference in the data. Thus, the team could compare new versus rebuilt equipment, automation versus manual, "live" versus computer-assisted help, and brand "A" versus brand "B."
- Environment: Differences are often the result of the environment under which or during which the product was produced or delivered. Do problems increase during the full moon? Stratifying the data by phases of the moon could lead to insight on the "full moon" syndrome. After defining what "inclement weather" means, stratify the data representing inclement days from the "fair weather" days to discover how inclement weather affects the process.
- Ability to Change: Lists of issues or alternatives often contain items over which one has no control, political agendas within the organization or directives required by law or regulation. Stratify a large list to pare down the list. Try to produce a data set containing data representing that which is possible to change.

Figure 5-3 includes two Run Charts. The top chart presents the customer complaint data representing customer complaints about "Product A." This chart suggests that complaints about the product have increased since Week Nine. The lower Run Chart stratifies the data by location.

Note that the second Run Chart in Figure 5-3, opposite, provides much "richer" information. Until Week 8, complaints were varying between zero and three both regions of the U.S. (though the East Region was running at a higher level). Then something occurred to increase complaints in the West Region while the East Region remained as before Week 8. The variability of the West Region complaint data has increased and complaints have increased. Variability, coupled with the higher number of complaints, would produce, not only a higher workload for customer service people, but also an unpredictable workload. This information is not evident in the combined data.

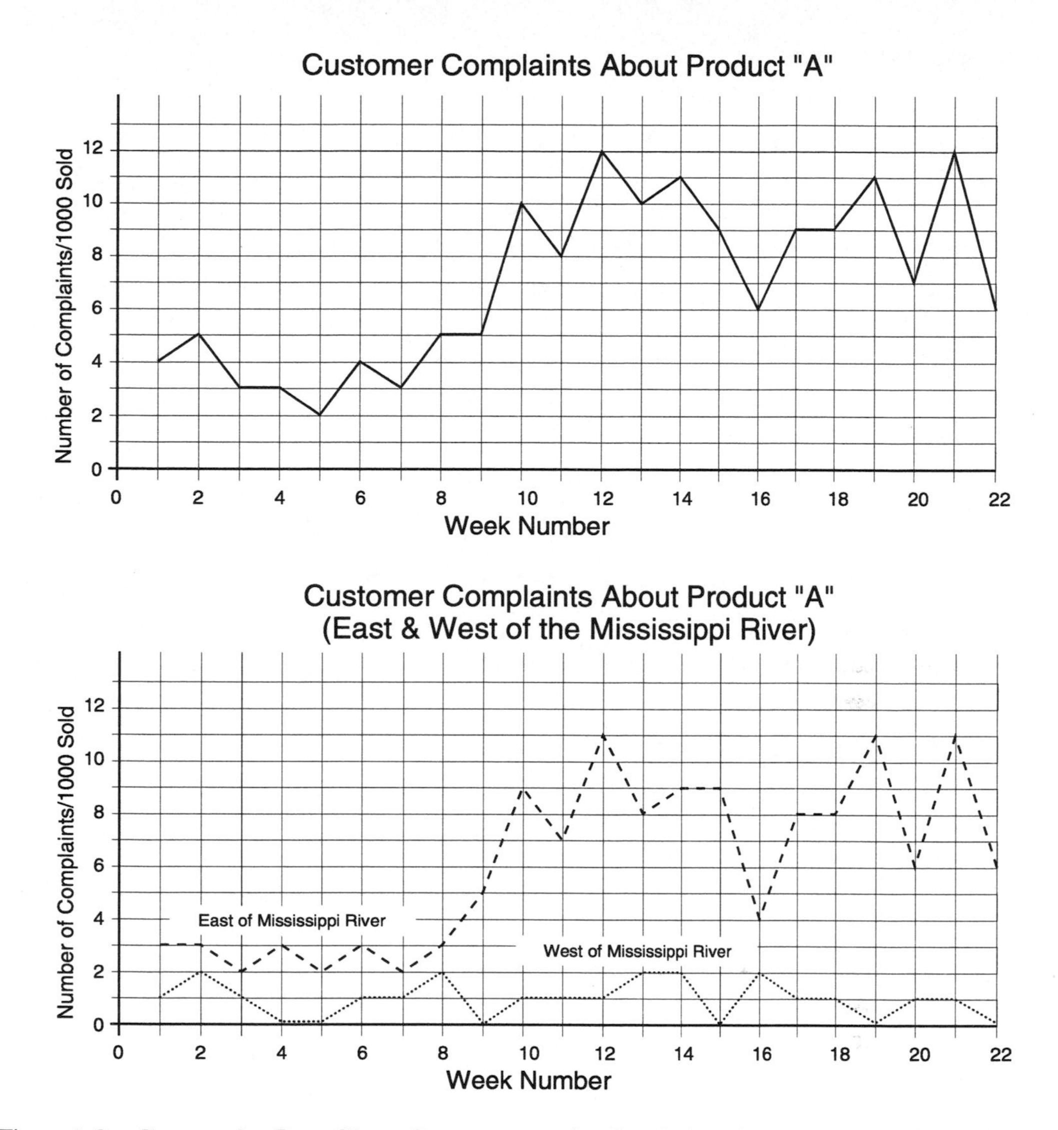

Figure 5-3 Composite Run Chart for customer feedback data (top chart) and Run Chart after data has been stratified by region of the U.S. (bottom chart.)

Matrix Diagrams

Middle and upper management have used the Matrix Diagram approach for many years for decision-making and project management. When used for these purposes, the Matrix Diagram appears complicated. A matrix, however, is simple to create and gives an individual or team a much better understanding of the information they have found or have generated. This provides a higher level of effectiveness as an individual or team endeavors to understand relationships among two sets of verbal data. A matrix also is beneficial when the time comes for making an informed, optimum decision.

A basic Matrix Diagram is a graphical method that visually presents the relationship between two or more variables. Organizations use the matrix approach for a variety of applications. As a tool for organizational improvement, the matrix approach will be limited to two variables[1]. The first variable is similar to an *independent* variable. It is the *given* variable or is the variable that drives the second variable. The second variable is similar to a *dependent* variable. It varies depending on what the first variable is. Intersections between the independent variables (rows) and the dependent variables (columns) possess a strength of relationship. Consider a simple three-by-three matrix diagram that helps in understanding what a customer might value and how the seller might satisfy each value. The matrix would appear as in Exhibit 5-1. This matrix shows the impact on the customer as a result of three different management strategies. In a real-life example, the matrix would probably become a seven-by-seven or even eight-by-eight. Actually, if a team were to survey the customer and brainstorm the possible independent variables related to filling customer needs, a 20-by-20 or greater matrix could result. It should be pointed out that a matrix could be three-by-four, eight-by-twelve, or any other resonable combination

1. Though this section confines the discussion of the Matrix Diagram approach to a two-variable matrix (sometimes referred to as the "L-Shaped Matrix"), three or more variables could be included. Examples of the three-variable matrix are the "Y-Shaped Matrix" and the "C-Shaped Matrix." Four variables could be matrixed using the "X-Shaped Matrix." If three or more variables must be matrixed, refer to "The Memory Jogger Plus+" by Michael Brassard.

Note: The information presented in this section is not meant to replace the other uses of the matrix approach; but, instead is meant to supplement other uses. Unfortunately, a matrix can become complex, quickly. This complexity leads to frustration. Frustration reduces the use of matrices to major projects or issues. This is unfortunate since the Matrix Diagram is a powerful tool and need not become complex to be highly useful.

After the matrix is completed with verbal data, give weights to the dependent variables (columns) to differentiate among the dependent variables. Convert "No," "Maybe," "Yes" and other alternatives to numerical values to make this work. The weighted numerical values for each row can be added and the one with the greatest point value is the "winner." Most people naturally do this exercise in their heads as they make decisions. The problem with a mental matrix is that the limit is usually a three-by-three matrix and converting matrix intersections to weighted point values is all but impossible.

Two types of matrix approaches are presented in this book. This chapter offers the Matrix Diagrams and the Synthesis Matrix, both useful in understanding the meaning behind verbal data and providing direction for improvement or change. Chapter Twelve presents the Decision Matrix. The Decision Matrix is quite useful as individuals or teams strive to make informed decisions, given two or more distinct decisions and two or more consequences of each decision.

Customer Expectations⇒ Cost Reduction Strategies ⇓	Low cost	Highest quality	Available within 24 hours
Buy supplies from lowest bidder	Yes	Probably not	Probably would not help
Maintain a very high inventory	No May increase	No relationship	Yes
Pay our employees the lowest wages possible	Yes	Probably not	Does not help

Exhibit 5-1 The basic, two variable, three-by-three matrix. Three cost reduction strategies are matrixed with three possible outcomes.

Synthesis Matrix

The Synthesis Matrix is an approach that individuals and teams can use to make much greater sense of the relationship among two sets of verbal data. Use a Synthesis Matrix to relate a set of characteristics and a set of options associated with those characteristics. The Synthesis Matrix is the basis for a much more complex approach: Quality Function Deployment (QFD). QFD is best left to those with a higher level of expertise in using matrices and is beyond the scope of this book.

The Synthesis Matrix will show the strengths of the relationships between any row and column. Numbers, symbols, words or letter grades suggest the strengths of the relationship. To this point, the Synthesis Matrix seems similar to the Decision Matrix. The primary difference is the use. Use the Decision Matrix to analyze alternatives and arrive at the best decision, given the situation. Use the Synthesis Matrix to synthesize information or put information together to provide direction.

Exhibit 5-2 shows a Synthesis Matrix used to transfer several customer focus groups' general expectations into the appropriate functions of the organization. The vertical column contains the general expectations of the customers. Across the top of the matrix are the functions or departments of the organization that could affect the various satisfaction factors. Give the intersections a score of high (H), medium (M), low (L) or none (left blank) to suggest to what extent the function or department could affect each satisfaction factor.

	Design	*Production*	*Marketing*	*Sales*	*Logistics*	*Accounts payable*	*After-sale support*	*Warranty admin.*
Initial quality	M	H	L	L	M	L	M	L
Suitability for use intended	M		H	H			L	
Reliabilty	H	M	L	L	L	L	H	L
Customer assistance			M	H	L	M	H	M
Product support meets needs			M	M	L	L	H	L
Cost-effective	H	M	L	M	L	L	M	M

Exhibit 5-2 Example of a Synthesis Matrix used to transfer general customer expectations into organization.

In the example shown in Exhibit 5-2, each function or department could then create its own Synthesis Matrix, initially focusing on the intersections with a "high" score. The design group would focus on how their designs affect reliability and cost-effectiveness. Not being certain how "cost-effectiveness" relates to product design, the design group starts with a Synthesis Matrix of this factor. List all general activities or steps involved in designing a product. List these activities or steps across the top of the matrix. The group then must determine what components make up the "cost-effectiveness" issue and list these down the left-hand side of the matrix. Returning to the raw data from the focus groups could help in completing this portion of the Synthesis Matrix. The group then fills the intersections with the strength designators. Now the design group has the necessary information to focus their efforts toward designs with higher levels of cost-effectiveness.

The Synthesis Matrix is most useful when striving to move information or concerns backwards or "up stream." This usually involves moving customer expectations or concerns back to the systems, processes or activities that have an impact on customer satisfaction. Internal to an organization, a Synthesis Matrix activity might be to matrix all types of benefits available to employees with characteristics of employees (age, seniority, type, etc.). Again, the high-strength intersections could lead to initiatives for benefit improvement.

Construction

Construction is straightforward. The individual or team must begin with the objective to translate information back into an organization, function, group, system, process or activity. Then, construct the Synthesis Matrix using following these steps:

1. Construct a matrix containing the necessary cells. A computer program, sheet of paper or white board will do. Teams will find the white board works best since modifying it during the exercise is easy. Different color Post-it® notes provide increased visual impact for the matrix intersections.
2. Determine the vertical column categories by obtaining the necessary information about the issue that will by synthesized. Call these the "dependent variables," since they depend upon the list developed in Step 3. Use surveys, questionnaires, customer response cards, complaints, focus group, and a variety of other data sources to obtain data. Try to narrow the number down to six to ten. Record these down the left side of the matrix.
3. List all the functions, departments, groups, processes, or activities that take place and affect the dependent variables listed in Step 2. Again, as in Step 2, try to keep these to six to ten items. Record these across the top of the matrix.
4. Determine the scoring approach. A common approach is the high, medium, low and none method.
5. Arrive at a score for each intersection that all team members accept. Individuals do not complete this part of the matrix as effectively as teams, unless focusing on processes for which they have complete responsibility.
   Optional: Shade the strong or intersections according to strength of relationship.
6. Find the strong relationships and begin determining how correction or improvement of the associated independent variable could produce a positive impact on the dependent variable.

The Matrix Diagram approach is also useful in making decisions. This use of a Matrix Diagram is presented in Chapter 12, "Decision Matrices."

Section II
Directing the Improvement Effort

Chapter 6

Project Selection, Prioritization, and Development

Selecting a Project

Projects can be selected in a variety of ways. Critical customers or increased pressure from competition may dictate the need for an improvement project. Changing environments can create projects. An analysis of what could go wrong can produce projects to reduce exposure to risk. An analysis of the cost of waste or non value-added activities often produce problem-solving or improvement projects.

Despite what has initiated the need for an improvement project, it is critical that a team chooses or receives a project that meets a variety of critical tests. The project must be manageable, yet stretch the team; within the collective expertise of the team; important, justifying the team's investment; consistent with the organization's strategic objectives; and possible within the time allowed. Other criteria are critical for newly formed teams made up of inexperienced team members. Project complexity, high profile and urgency will reduce a newer team's opportunity to become effective. Most teams with projects aimed at making changes surrounding "political" issues will fail or will address the "edges" of the issue, only.

Finally, this section offers a cost/benefit analysis, useful as projects are selected and as an individual or team sorts out solution or improvement options.

Failure Mode and Effect Analysis

One way to select an important project is to perform a Failure Mode and Effect analysis. Failure Mode and Effect Analysis (FMEA) is an organized approach to determining:

1. Possible component failures (and, therefore, product failures),
2. How it goes about failing (failure mode),
3. How critical the failure will be (effect analysis), and
4. How to reduce or eliminate the possibility of the failure (purpose of FMEA).

FMEA is a six-step approach and is most powerful when utilized by a team.

Failure Mode and Effects Analysis Form

Critical Component	Possible failure	Cause of failure	P x D x S =T*				Effect of failure	How can the failure be eliminated or reduce?
			P	D	S	T		

* "P" = Probability of Occurrence, "D" = Damage Likely, "S" = Severity of the Damage and the Consequences, and "T" = Total Score for the Hazard. Scoring : "1" = Low to "5" = High.

Exhibit 6-1 Failure Mode and Effect Analysis Form.

Procedure:

Step 1: Identify all the critical components, or collection of components, making up a product or critical subassembly, and then, determine which components are:

- Likely to cause serious product failure (or degrade performance considerably) if they fail?
- Likely to cause other components to fail?
- Single (sole) source items?
- Limited life or high wear items?
- Difficult to maintain or adjust?
- Difficult to replace?

These ones meeting one or more of these criteria make the FMEA list.

Step 2: What, exactly, is the failure mode for each component or collection of components on the FMEA list?

- How is the failure initiated?
- What is the failure mode or the reaction of component to failure of the mechanism?

Step 3: What causes the failure?

Step 4: How serious is the failure? (1 = very low to 5 = very high)

- (P) What is the probability of occurrence?
- (D) What is the likelihood of damage to the product or subassembly?
- (S) To what extent will serious damage result (concerning cost to correct, damage to other components, and injury to the user or others)?

Multiply P times D times S to get a seriousness of failure score, "T." Typically, a score of nine or more suggests a level of seriousness high enough to be a candidate for Step Six.

Step 5: What is the effect of the failure?

- How is the failure detected?
- At what point is the failure detected?
- What is the immediate effect?
- What is the ultimate effect?
- How is system performance affected?
- Does failure require an immediate repair or is repair possible during an off cycle?
- Is it necessary to remove something for repair to or replacement of the failed component? What?
- What special tools are necessary to correct the failure?
- What can be done to eliminate or at least dramatically reduce the failure?
- Design the component out of the system or design a different component into the system
- Require strict conformance to standards during manufacture, assembly and test
- Improve the process for producing and using the component
- Failsafe the manufacture, assembly and test
- Develop and implement appropriate training for installation, use and repair of the system
- Develop and implement strict preventive or predictive maintenance requirements
- Develop and implement strict procedures for field test, adjustment and repair

Step 6: Assign responsibility for improvement projects suggested by Step 5. These projects could be appropriate for

- Individual experts or expert teams
- Managed project teams with a project manager

Prioritizing Hazard Reduction Probability/Severity/Cost Approach

Everyday activities expose all organizations to hazards. Single sourcing of a critical supply without developing a partnership with the supplier, and a contingency plan, produces exposure to a hazard. The supplier could experience any number of problems delivering supplies, some beyond their control. A flood, snowstorm or earthquake could halt tomorrow's delivery of needed materials in a just-in-time strategy. Key employees without backups expose an organization to a hazard. Introducing a product or eliminating a product, prematurely, could be a hazardous situation. Extending preventive maintenance period on a critical piece of equipment could produce a hazard.

How does an organization deal with all potential hazards, understanding that eliminating all hazards will be impossible? Since reducing or eliminating all exposure to hazards is impossible, the organization and groups within the organization must use a proven approach to ranking the hazards. The Probability/Severity/Cost or PSC approach will help. The PSC approach will surface the hazards that require reduction or elimination, now. A PSC analysis should be done regularly to bring the next phase of hazard reduction to the surface and identify new hazards worth concern.

The activity of setting priorities to highlight which hazard exposures among many require attention, a person or team must consider three important criteria:

1. Probability of occurrence;
2. Severity of the effects to the organization, employees, customers or the community; and
3. Cost to prevent future occurrences, reduce the severity of each occurrence, or reduce the number of occurrences.

Assign each criteria for a failure mode an importance rating, as designated by a "grade." An "A" represents high severity, a high probability of occurrence or a low cost (investment) to reduce drastically or prevent the hazard. An "F" represents a failure mode resulting in very little severity, or one that is unlikely to occur or one costing a great deal of investment to prevent or reduce. Therefore, a triple A (AAA) hazard is a primary candidate for immediate action. A triple A hazard could produce severe outcomes should the hazard become a reality, is highly likely and does not cost that much to prevent or drastically reduce. Ignore an "FFF" hazard.

The next step in hazard prioritization is to assign numerical values to the grades. "A" would equal 5, "B" would equal 4, "C" would equal 3, "D" would equal 2 and "F" would equal 1. Then, multiply the three values to get a hazard score. A triple A (5x5x5) totals 125 and a triple F (1x1x1) would total one. Multiply each individual criterion score to produce more discrimination between hazards. Those hazards requiring immediate attention become obvious. Exhibit 6-1shows an example of the various scores for each of six different hazards.

One additional scoring scheme may be useful in unique situations. Often weighing one or two of the criteria to reflect business conditions realistically is appropriate. Perhaps an organization is operating with limited resources. Enhance the effect of high cost to direct the maximum hazard reduction efforts with the limited resources available. Give the grades for the cost factor of each less weight to reduce the chances of higher cost, higher probability, higher severity hazards overpowering lower cost alternative projects. Perhaps "A" = 5, "B" = 3, "C" = 1, "D" = 0 and "F" = 0 would present a more accurate picture of each hazard as to cost constraints (as noted in the "Cost Weighted" column). Exhibit 6-2 depicts how cost-weighed scores cause the second hazard (# 2) to become the prime candidate because it is less expensive to reduce or eliminate.

Hazard	Grades	Score	Cost Weighted
#1	AAC	5x5x3 = 75	5x5x1 = 25
#2	ADA	5x2x5 = 50	5x2x5 = 50
#3	ABD	5x4x2 = 40	5x4x0 = 0
#4	CBC	3x4x3 = 36	3x4x1 = 12
#5	ACF	5x3x1 = 15	5x3x0 = 0
#6	FFC	1x1x3 = 3	1x1x1 = 1

Exhibit 6-2 A hazard prioritization example.

Note the change in the rankings when cost is weighted. The first and second hazards still warrant immediate attention, but the fourth hazard becomes the next project.

Team Approach to Hazard Prioritization

The PSC approach is an excellent team tool. Since teams consume more resources, it makes sense for the team to concentrate efforts on those hazards that have the greatest probability of occurrence, could produce the most severe effects on the organization and are cost-effective to reduce or eliminate.

Procedure:

1. Decide upon a grading or scoring system to apply to the analysis, grades A through F. For point values, five through one are possibilities. If money is limited, weigh the cost scores as in the individual PSC.
2. Brainstorm a list of all possible hazards within the span of control or concern of the team. See Chapter 9 for Brainstorming. Do not discount or discard any hazards at this point. All are worth listing and sorting will take place later.
3. Discuss all hazards to attain understanding by all team members. Try to reduce lobbying for "favorite" hazards.
4. Assign A and F scores in each criteria for each hazard, as warranted. Return to the unscored criteria and assign B and D scores. Hazard criteria not scored at this point get a C. Start with the probability of occurrence, then move to severity and finally to anticipate cost to prevent or reduce the hazard. The team must remember that this is a rough pass to surface those hazards worthy of team or individual attention.
5. Convert the grades to numerical scores and multiply each hazard's three scores. This produces a hazard reduction priority list.
6. For those groups charged with selecting a hazard, the top scoring hazard is a likely candidate. Some hazards on the list could be candidates for internal or external experts. The team could make recommendations accordingly. Assign some highest ranked hazards to other hazard reduction teams.

Cost of Non-Quality Analysis
Using the Cost of Quality to Select and Justify Projects

Most organizations plan and track activities in financial terms. For-profit organizations, in particular, manage for month-end, quarter-end and/or year-end results. This is entirely appropriate. Money is a major input and output for these organizations. Money is a measure of organizational success or failure. Finances drive the non-profits and institutions also. Money comes into the organization from taxes, donations or grants. Rather than profit, the organization strives to contain spending equal to funds available. Why is it then that financial considerations always appear to be a minor part of the improvement effort and a major part of cost reduction efforts?

The problem is that two extremely important financial categories are usually missing from the financial data of most organizations. Many aspects of the cost of non-quality and the financial payoffs of quality improvement efforts are, for the most part, not recorded, at least with any degree of accuracy. The rule-of-thumb has been simply to believe that anything done to improve processes within an organization, will improve the financial results or use of funds. This belief requires a great deal of faith and, too often, patience waiting for the payback. Many organizations have found improvement efforts ebbing due to the lack of a clear, documented financial payback. Some of these organizations have renewed their initiative by learning and applying a technique know as "Return on Quality"or more accurately stated, "Return on Improvement Investment" (ROII). ROII also provides data to rank improvement projects. ROII, as a decision-making mechanism, is discussed in Chapter 12.

Return on improvement investment consists of two primary financial measures. The first is the total cost to improve products, systems, processes and activities. The second is the reduction of a major component of ROII: the cost of quality (COQ). COQ documents investments in good quality and costs resulting from non-quality. COQ has four components:

1. Prevention: The investments made to assure that a process or product will deliver what the customer requires, including activities to identify and eliminate root causes of problems,
2. Appraisal: The investments made to detect problems or poor quality before the goods or services get to the customer,
3. Internal Failures: The costs associated with activities that correct a problem or defect after it occurs, but before delivery to the external customer, and

4. External Failures: The costs associated with activities that correct a problem or defect after it occurs upon or after deliver to the external customer.

Together, these four can account for 20 to 50% of an organization's income. The critical issue is how the organization goes about optimizing the COQ investment by:

1. Driving the failure costs to zero,
2. Reducing appraisal costs since these costs do little to add value for the customer, and
3. Maximize the value received from prevention and improvement investments.

ROII tracks maximization of prevention and improvement investments. Any prevention effort costing $10,000 should provide a payback of more than $10,000 within an appropriate period. The same is true of improvement initiatives. Determining what it costs to discover and implement improvement is not the problem. The problem is how to put a dollar figure on the results and setting a payback period. The traditional one-year or 18 month payback period may be inappropriate. Also, since many other internal and external factors affect an organization, there will be difficulty attributing monetary gains to a specific improvement effort.

Despite the problems and inaccuracies, individuals and organizations must track the costs and returns associated with improvement efforts in the same manner as all other costs and returns. The difference is that the costs are not accurate, and may even be guesses. That is all right, the "improvement financials" are indicators, not part of the profit and loss statement or a report to a funding agency.

Improvement Investments

Improvements generally come at a cost. Most of these costs can be found in a variety of accounting reports, but specific improvement projects probably do not get these charges. These costs include:

- Time away from the job for individuals engaged in improvement activities and team meetings (wages, benefits, overtime, etc.)
- Surveys, focus groups and other data-gathering activities
- Supplies used for meetings and to test improvements
- Costs for Design of Experiments (DOE)
- Consultants and experts
- Payments to individuals for acceptable suggestions and other recognition costs

- Space for meeting rooms and other improvement activities
- Costs associated with pilots and tests
- Reduced efficiency during achievement of the improvement
- Costs to document, communicate, educate and train

Prevention Investments

No one can argue that prevention is the preferred approach to improving quality. However, one could argue about what makes up prevention cost categories and how to figure out the costs of these categories. Fortunately, prevention costs are easy to determine, once one identifies the categories. What is more difficult to determine are the costs resulting from the lack of prevention. The various prevention costs include:

- Costs similar to improvement activities, plus
- The use of prevention approaches such as quality function deployment (QFD), failure mode and effect analysis (FMEA) and capability studies
- New employee orientation
- Employee, customer and supplier education
- Product (goods and services) and design reviews
- Supplier evaluations
- Preventive maintenance
- Pilot studies
- System and process capability studies and control
- Mistake-proofing
- Improved communication and feedback processes

Obviously, prevention requires a variety of investments. When a problem occurs, it is too late to prevent the problem. Though the customer (internal or external) may subsequently become satisfied after defects and disappointments are corrected, the costs of correction after delivery are high.

Return on Improvement Investment

One activity an improvement or problem-solving initiative should include is a cost/benefit analysis. At an early point in the project, when the scope of the project is apparent, the individual or team should consider the cost to develop and carry out the improvement. Also, determine the savings resulting from the improvement. The savings should also include other revenue improvements resulting from the improvement. Time of the payback should be considered, also. Many organizations use a rule-of-thumb of one year for a payback. If the investment cannot result in a savings during the first 12 months after implementation greater than the cost to develop and implement, question the funding of the project. There is no "magic" in a 12-month payback. Some improvement projects may take two or three years for a payback. The decision to fund a project depends on other factors. These factors include how long the payback will continue beyond the initial investment payback and to what extent the improvement strengthens the organization. Also, remember that one successful improvement may spawn other improvements. This is why the seventh step in the Systematic Problem-Solving and Process Improvement Process (Chapter Ten) is critical. The seventh step diffuses learning and success from one project to other projects. Do not forget to consider the other returns of a project such as the value of learning, if learning will be diffused to other projects.

It is not as critical to detect these costs and returns accurately. It is critical that they are considered and estimated. These figures are not used for traditional accounting and finance purposes such as tax computations, profit and lost results, or to calculate bonuses. The uses of these costs are to determine the importance of a project and to estimate how much time and money to invest in the improvement efforts. For example, if an investment of $500 to $800 will result in a one-year payback of $3,000 to $6,000, the improvement effort is worthwhile, though the cost and savings figures are rough estimates.

The results of the improvement cost/benefit analysis will place the improvement project in one of three categories:

1. The project cannot be justified because the investment outweighs the benefit. The individual or team should move on to another project with a positive return on investment.
2. If the project is considered absolutely necessary and the return on improvement investment is negative or minimally positive, then the organization must fund the project without consideration of a positive return on investment. A more attractive alternative is to find ways to reduce the investment required while improving the return from the investment.
3. Projects with outcome returns outweighing investments should continue immediately, assuming the organization currently has the resources necessary. Do not forget that resources include people's time, expertise, facilities, and money.

The organization should use three categories of costs when considering cost reduction because of an improvement or problem-solving project. Appraisal and inspection require resources to attain the required levels of quality. Some appraisal costs are investments incurred before standardizing processes. These are really prevention investments. These are investments to make certain the process will produce quality over the long term. Failure costs are always undesirable costs, no matter where they occur. Internal failure costs are usually less and affect the customer less, but can become massive if processes are out of control. The external customers notice external failure costs. External failure costs often produce side effects such as loss of good will and negative publicity. These costs are extremely difficult to measure, but can be very high.

Appraisal Costs

Appraisal costs include all those costs associated with assuring that the process meets customer requirements, quality systems and process improvements are in place and remain in place, and random inspection proves the process continues to provide conforming output. Appraisal activities actually consist of two components, audits to confirm quality will continue, and inspection activities to find problems. Audits are proactive, reducing the chance of non-quality outputs. Inspection is reactive, ferreting out non-quality, after the fact. Appraisal costs include:

- Prototype inspection and test
- Product specification confirmation analysis (prevention)
- Quality audits (prevention)
- Receiving inspection and test
- Proofreading
- Customer, employee and supplier surveys (mostly prevention)

Audit and inspection activities will increase during extensive improvement periods. Successful process improvement will virtually eliminate the inspection costs but will only reduce the audit costs. This is why reporting audit and inspection costs as separate components is best. If the process is perfect, then the appraisal cost becomes inspection rather than an audit cost. An example of inspection is receiving inspection and proof-reading. These activities are of little value in a perfect organization or process.

Internal Failure Costs

Internal failure is one aspect of the cost of quality. Most organizations identify and document some of these costs. Internal failure costs consist of:

- Scrap and spoilage
- Disposal of waste due to problems
- Rework and doing things over
- Overtime to make up for problems or failures
- Production downtime or delayed deliveries
- Cost of changes due to mistakes

Other costs of internal failures are more difficult to figure out, and some are not only difficult to identify, but are near impossible to determine:

- Turnover due to management systems and approaches producing dissatisfied employees
- Delays and idled employees
- Problem investigation costs
- Revisions and changes due to problems
- Excessive inventory carrying costs (just-in-case inventory)
- Fines, legal actions and judgments
- Lost customers and lost opportunities
- Employee dissatisfaction

Rule of Thumb: If the cost would go away or become unnecessary if all processes were perfect, it is a failure cost. Remember: This also includes all management processes. Management processes are the processes of managing people and leading the organization.

External Failure Costs

As with internal failures, some external failures are easy to document and report. These include:

- Warranty and out-of-warranty costs
- Adjustments and special discounts
- Complaint handling systems
- Product recall and liability costs
- Claims, legal actions and awards
- Giveaways (to mollify a customer or stave a legal action)

A variety of external failures are difficult to identify and cost. Therefore, the imperative becomes to find, estimate and include these costs in the ROII calculation.
These costs include:

- Customer and potential customer bad will
- Lost customers
- Disappointing customers trying to reduce warranty costs
- Overdue accounts due to product-related problems
- Lost opportunities
- Employee dissatisfaction

The Cost of Non-Quality provides baseline and follow-up data to check the effectiveness of improvement efforts. Just remember that as the organization improves its ability to track and determine the Cost of Non-Quality, these cost figures will naturally increase. Members of the organization must not become discouraged. Opportunities for improvement become apparent as the organization uncovers the Cost of Non-Quality.

The "Project Analysis and Selection Form" shown in Exhibit 6-3, helps a team in selecting an improvement project. Discussing each factor is the first benefit of the form. Members will gain considerable insight into the project because of the selection criteria discussion. The second benefit occurs as the team members arrive at a consensus on the team vote and each criteria's weight. Divide the sum of the weighted scores by the maximum possible sum of weighted scores. Multiply the result by 100 to estimate the team's chance of success, not given special circumstances.

Teams with several project options or teams that have broken the project into several more manageable projects should use the "Project Analysis and Selection" form to help the team in choosing the best project for the situation. The team should use the weighing feature to focus the team on those criteria that will have the greatest effect on success or produce the greatest improvement.

Project Analysis and Selection

Proper project selection is critical to the success of a team, particularly a newly formed team. Occasionally the organization will assign a project to a team beyond the scope or expertise of the team members. The project may be too massive or may contain many components. Many other criteria predict the success or failure of an improvement project. This is why the selection and focus process is critical to the team's success.

Project Analysis and Selection Form

Project: ______________________ Date: __________

Team: ______________________

Criteria:	Comment	Team Vote	Criteria Weight	Weighted Score
Chances for Success (Really)				
Scope (The narrower, the better)*				
Cycle Time (Frequency of output)*				
Significance (Really important?)				
Cost Savings (Worthy of team's)				
High profile (Visible?)				
Feasibility With Allocated				
Not Too Complex (Break it down?)*				
Possibility Within Given Time Span*				
Not Political (Hinders a project)*				
People Generally Care It*				
Process Owners Want Imrovement*				
Impacts External Customers				
Urgency (OK, if adequate resources)				
No (or Little) Resistance to Change*				
Consistent With Strategic Objectives*				
Existing Process*				
		Totals	100	

* These criteria are critical for new teams.

Exhibit 6-3 Project Analysis and Selection Form.

Procedure:

1. The team, by consensus vote, determines the weight assignments for each criterion by dividing 100 points among the criteria. A weight of zero is possible, suggesting that the criterion does not apply or is unimportant. Place the consensus weight for each criterion in the "Criteria Weight" column.
2. Discuss each criterion and critical comments. Note comments directly on the form or by using letters to refer to comments recorded on a separate page. Keeping a record of critical comments during this discussion is important. Use "Judging Each Project Selection Factor" which follows this procedure.
3. After the discussion of each criterion, the team takes a consensus vote. A quick method is to ask for a "Thumbs Up" vote. See Chapter 12, "Optimizing Decisions." Give thumbs up a three, thumbs sideways a one and thumbs down a zero. Place the total in the "Team Vote" column.
4. When the list is completed, multiply each criterion's score by the weight to arrive at the "Weighted Score."
5. Total the "Weighted Scores" to arrive at a total score for the project.
6. If other projects remain for analysis, move the next potential project and go through above steps. Otherwise, move to Step 8.
7. After scoring all the potential projects, consider the project with the highest score. If the top two or three scores are within a narrow range, return to the matrix for each and discuss those criteria with lower team votes. Also, revisit all criteria weights. Are the weights consistent with the team's maturity and status within the organization? Are the weights distributed so that it is clear what is most important? Remember: Discussion of the criteria is as important as the scoring process.
8. When the field of possible projects is narrowed to one, or the team is dealing with one project, examine those weighted scores that are lower than anticipated. Note what precautions the team must take. Consider these questions and issues:
   - Is the project really doable? If the project is not doable, what would make it more doable?
   - Is the project's scope too wide? How could it be narrowed? Is there any way to break the project into several projects?
   - Usually, not much can be done to decrease cycle time. However, is there some way for the team to test a solution or improvement in a mid cycle?
   - Significant and high profile projects must be completed on time and correctly. Planning is critical. Is the project too significant or the profile too high for a newer or lower level team? Does the team have the necessary management support?

- Is there adequate funding for this project? Inadequate funding will produce problems. Identify this situation when possible and take steps to assure adequate funding.
- Is the project too complex for this team? Complex projects produce high failure rates for newer teams and can even overwhelm a mature team.
- What is the "political" nature of this project? Will the project touch forbidden turf? If the project is political, the team must be very careful, even to the point of avoiding the project.
- How do the process owners feel about the project? If people of the organization and the process owners care about the project, it has a much higher chance of success.
- To what extent will this project impact external or ultimate customers? Carefully execute projects that impact external customers.
- How urgent is this project? Is the urgency or lack of urgency imagined? Urgency is an interesting issue. If urgency does not exist, team members may not commit the time or energy necessary to see the project to swift completion. With a sense of high urgency, the team may hurry the project, installing temporary fixes rather than taking the time to find and implement solutions.
- Uncover all possible resistance to change. With resistance to change, the best solutions may suffer from poor implementation. Reduce resistance by including process owners and process customers on the team. Keeping the organization informed is critical.
- Are the strategic objectives that affect this project known? Projects that impact the organization must be consistent with the strategic objectives of the organization. This may seem like an obvious statement, but too often the best intentions may not align with the strategic objectives. Localized projects may require a test against those strategic objectives that apply to the project's scope.
- Finally, does this project concern an existing process in need of problem solving or one that would benefit from improvement? Existing processes are much better for newer teams. Existing processes provide current situation data, identified process operators and customers. Newer teams should avoid taking on a project that creates a new process, unless the members are from mature teams that have successfully completed other projects, including new process projects.

Judging Each Project Selection Factor

Reasonable Chance for Success:

Early projects, particularly the team's first project, must have high chances of success. In fact, the first project should be a certain winner, even if it means selecting a minor project. For newer teams, give this criterion heavier weight. Permit mature teams to tackle projects with lower chances of success, especially if the project is critical and urgent. Excellent planning will be necessary.

Narrow Scope:

The scope of a project must be kept quite narrow for early team projects and reasonablely narrow for later projects. Err on the side of more narrow than wider. Teams function much better if they can remain focused. If the scope is wide, break the project into smaller projects or into its components. Individuals also benefit from a project with a narrow scope.

Short Cycle Time:

The time from initiation of the process to identifiable output is the cycle time. A cycle time of six months requires a wait of six months from implementation to detect improvement results. The weekly payroll is an example of a seven-day cycle time. Any improvement in payroll activities should be evident after a week. Most organizations conduct budgeting annually and any improvement in the budgeting system will not be fully evident until a year later. Remember that it is desirable to measure at least 20 outputs (samples) to be minimally certain of the ongoing process output.

Significant:

The project must be worth the time, as measured by money, affect on employees and customers, future problems, etc. All projects requiring a team must be important to the supervisors or managers of team members because the supervisors must clear the way for team members' time. Supervisors must view team membership as part of each team member's job to keep the member's assigned job from interfering with team membership. The team members must perceive the project as important so that the team allots time and energy to the project.

Produces Cost Savings:

The potential cost savings over a period is a critical measure of what management and the team will commit to the project. Traditionally, the period is 12 months, but a longer period can be chosen. Will the cost of solving the problem or improving the process outweigh the savings derived during the payback period? If so, the project may be destined for failure.

Profile:

The higher the profile of the project the more likely the team members will commit to the project and outside resources will be available. However, high profile projects may run into problems early in the process if team members or others perceive that the project is moving slowly, or if the setbacks are communicated better than progress.

Lower Complexity:

Projects of higher complexity are much more likely to result in failure, especially the earlier projects in the initial stages of "quality management" implementation. Weigh this heavy for the first one or two projects a team considers. Complex projects are best broken down into several more manageable projects. Judge the less complex projects using the Project Analysis and Selection Form to zero in on the best choice for the first project.

Feasibility With Allocated Resources:

If the allocated resources (members' time away from their regular job, time of other people, meeting space, materials, supplies, equipment, money, etc.) fall short of what the team requires to complete the project, the likelihood of failure is much higher. A team should carefully consider avoiding obviously under-funded projects, or make it clear to the person or group assigning the project that the team probably cannot implement an optimum solution.

Possibility Within Given Time Span:

If completion of the project requires more time than available, the likelihood of a failure or team member dropout is much higher. Make certain that adequate time is available. The exceptions are projects with high urgency. Generally, not enough time is available to complete urgent projects to the satisfaction of all team members.

Not Political:

Political issues make extremely difficult projects. Save these for later or suggest use of a different mechanism. (Examples of political issues are parking spaces, office sizes and locations, and pay mechanisms, etc.) Projects of a political nature are often best left to teams from higher level of the organization. Just make certain to gather information from those affected by the project. All projects associated with labor union contractual matters should be left for union-management mechanisms.

Similar to political issues are those projects concerning that which the organization holds "sacred." Elimination of very low margin customers who have been customers for many years is an example. Others might be employees' titles or how people furnish their offices.

People Generally Care About It:

People generally caring about the solution or improvement greatly increases the likelihood of success. New teams should focus on projects about which people care.

Process Owners Want Improvement:

If the process owners (operators) value the possibility of problem elimination or improvement, success will be much more likely. Do not forget to include process owners on the team, if possible.

Impacts External Customers:

If the project impacts external customers, the project will have a much higher visibility and importance. More resources will probably be available. Take greater care to insure team success if the project affects external customers. Also, the team should be more mature, with a higher level of expertise.

Urgency:

Urgency increases the chances that meetings will take place on a more regular basis, more often, with greater progress. Urgency also keeps the project on schedule since management will not forget a team is addressing the problem or opportunity. However, urgency can cause the team to take shortcuts or carry out quick fixes rather than long-lasting solutions. (Perhaps a PSC analysis should take place.)

Lower Resistance to Change:

Resistance to change is the leading killer of solutions and improvement. Consider this element very carefully. If resistance is certain to be present, plan on how to overcome or dramatically reduce resistance.

Consistent With Strategic Objectives:

Projects must align with strategic objectives of the organization, and the group owning the process being improved. Otherwise, people give the project secondary importance and the team may fade away. A project not consistent with strategic objectives wastes time and resources since implementation is generally impossible. Worse yet, another team may have to dismantle a successfully implemented change.

Existing Process:

If the process is an existing process, the team has the advantage of existing customers and existing data. This is a plus. If the team's project concerns a new process, customers and data will be much more difficult or impossible to obtain. Consider this when voting on this criteria. Existing processes should be candidates for problem-solving or improvement early in the "quality management" implementation process. Newly formed teams with inexperienced members should focus on problem-solving projects. New processes, systems and products are best left to "managed" projects. Managed projects benefit from an assigned project manager, one who can commit larger amounts of time and has expertise in managing this type project.

Some projects are completely outside the realm of improvement teams. If the project concerns any of these issues, hand it off to another type team, as noted. Projects to avoid are:

- Union contractual issues or any issue that falls within the domain of union-management negotiations or general concern. Defer these to the union-management team.
- Global issues. If a project will result in a dramatic improvement in the organization, suspect the project as a global project. Return to the "Narrow Scope" and "Lower Complexity" selection factors or defer the more global project to a much higher level, elite team.
- Impossible issues. The improvement team should not attempt to undertake projects to change laws or regulations, attain 100 percent attendance every day, improve a good or service to the point it will satisfy all the customers all the time or eliminate policies and procedures. If any of these projects become critical to the organization, an elite, high level team should take it on.

Cost/Benefit Analysis

Many decisions come down to determining the benefit and cost of each alternative. Often, lower cost alternatives produce smaller benefits while higher cost alternatives produce larger benefits. If limited monetary resources are available, the decision is simple: Go with the lower cost alternative. However, how does a team or individual know if more money is available? How does the team determine that a higher investment is justified? Additionally, how are higher-cost alternatives without proportionally higher benefits separated out? This is the purpose of the "Cost/Benefit Analysis Form" shown in Exhibit 6-4.

Cost/Benefit Analysis Form

Possible Solutions/Improvements	Chances of Success With Allocated Resources	Total Cost to Implement Solution or Improvement	Benefits		Cost to Benefit Ratio	Payback		Rank of Option
			Short-term	Long-term		Payback Delay	Payback Duration	
First option:								
Second option:								
Third option:								
Fourth option:								
Fifth option:								
Sixth option:								
Examples of possible Measures:	Assured High Medium Low Unlikely Uncertain	Actual: $ 1,250 Range: $1,000-3,000 Estimate: $5,000+	Actual $ or estimate Quantitative statement: Uncertain One-year payback High, Medium or low Increased sales Improved customer satisfaction Lower costs Less rework		Very high High Medium Low Very Low None Negative	Number of months or years Immediate Short Long Uncertain	Number of months or years Short Long Forever Uncertain Until process product or environment changes	First second, third, etc. Very High, high medium or low

Exhibit 6-4 Cost/Benefit Analysis form.

Procedure:

1. List the potential solutions or improvements in the space provided. Avoid analyzing more than four options even though there is space for six options. Use other approaches to narrow the possibilities to four or less.
2. Next, decide what chances for success exist for each option. Consider all factors affecting the success of the option listed in the Project Analysis and Selection matrix in this section.
3. Add all costs to implement the improvement or solution for each option.
4. Itemize the benefits of each option as either short-term (possibly one year or less) and long-term (greater than one year). A variety of ways exist to list the benefits, including actual cost savings or increased sales.
5. Computing an actual cost/benefit ratio is best, but this is usually not possible. An individual may find it difficult to come up with a qualitative statement that is easy to defend. A team can choose a qualitative statement through a consensus approach and better defend the decision.
6. Another critical aspect of the decision is the delay between decision (or implementation) and the beginning of the payback. Those options that have a higher cost/benefit ratio and a shorter payback delay are better choices.
7. The duration of the payback is also important. If the cost benefit ratio is high but the duration is short, the option may not be the best choice among the options. A much longer payback duration from a lower cost/benefit ratio might be a better choice.
8. The last column compiles the information from the other columns. For an individual, this column forms the value judgement for each row. For a team, this column becomes a consensus decision. The advantage of a team becomes apparent as the team members refer themselves to information in the columns to optimize the statement in this column.

Chapter 7

Four Improvement Strategies: Using the Four "Res" to Select Your Improvement Approaches

Improvement Strategies

Improvement can result from four uniquely different approaches or strategies. If a product, system, process or activity is experiencing a problem, an individual or team must solve the problem. Lacking problems, further improvement might be considered to improve systems, processes and activities, continually. Often, simple continuous improvement in small doses just does not result in the required improvement. Then, approaches or techniques that provide a quantum leap improvement should be considered. Frequently, significant improvement results from the application of advances in technology or highly creative thinking. Occasionally, improving what is current will just not result in enough improvement to produce customer satisfaction, let alone delight, or will not result in levels of efficiency to assure survival of an organization. The last resort is to completely forget the current approach and rethink the system, process or activity, beginning with customer needs and expectations. This is reinvention; also, called "reengineering."

Individuals and teams should use these four improvement strategies in concert within the organization. Few organizations are void of problems, lack opportunities for continuous improvement, cannot benefit from some innovation and breakthrough, and exist in an environment void of pressures for dramatic change.

Improvement Strategies: The Four "Res"

Improvement experts often recite the proverb, "If you only have a hammer, everything looks like a nail." This represents what has happened in the recent history of quality improvement. Often, a single improvement strategy gains such favor in an organization that it becomes the strategy of choice, almost to the exclusion of all other approaches. This must not be. A variety of approaches, techniques and tools are available for improving systems, processes, and activities that produce goods or services for groups or individuals. A hammer may be the best choice for a tool for a specific project, but the toolbox contains many other tools.

The toolbox for improvement activities contains four sections. These are the four improvement strategies: Repair, Refine, Renovate and Reinvent . . . the four "Res" of organizational improvement.

"Repair" occurs at two levels. The first level includes those activities involving the execution of quick fixes, shoring up weak areas. Many call this the "band aid" strategy. The second Repair strategy goes deeper, removing the root cause of problems to prevent their return. "Refine" involves those activities associated with continually improving that which is not broken: The products, systems, processes and activities that make up the organization. "Renovate," as the word would infer, includes the efforts that result in major improvement, going beyond repair and refinement. The results of renovation are usually dramatic. "Reinvent" is the fourth, and highest level of improvement strategies. Reinvention requires the team or individual to forget the current approach. Starting with a clean sheet of paper and a clear understanding of customer requirements, the goal is to develop a new product, system, process or activity.

Consider building a new house. A new house is seldom perfect and this why the contractor expects the owner to keep a "punch list." Items on the punch list require repair. The house may be perfect when the family finally moves in, but will not remain perfect. Items age. Things break down. To keep the entire house functioning, occasional repair is necessary. What was perfect on paper is not quite perfect in reality. Many small things could be better. This calls for refinement. Many small improvements over considerable time make up refinement. After many years in a house, families find that the way to make the house fit the family needs is to renovate. Perhaps an addition to add another bedroom, modernize the kitchen and install a high efficiency furnace. Occasionally, even extensive renovation will not meet the changing needs of the family. The family needs to rethink their home needs. The family must decide which current needs are critical, which future needs to consider now and what resources are available. Chances are very high that the replacement house will be completely different. The family builds or purchases the new house and the process of repair, refine and renovate begins.

Use these four basic strategies to improve:

- Products (the goods or services that a system or macro process produces),
- Systems (a collection of processes producing an output that has major impact on the organization or external customers),
- Process (a collection of activities that transforms inputs, adding value to an output), and
- Activities (the observable actions making up a process; work).

Understand each strategy and use each at the right time for the right situation.

The Four Res

Knowing the intent and components of several improvement strategies leads one to the expanded fable: "Given a kit of tools, everything looks like an opportunity." The four improvement strategies are quite different. Repair (fix) what is broken. Refine (improve) the unbroken to make better. Refurbish (apply innovation and creative thinking) the unbroken which require quantum improvement. Finally, Reinvent (completely rethink) that which would never satisfy the customer, whether internal or external, despite the improvement. Coexistence of the four strategies is not only possible, but absolutely necessary, in all organizations. Nevertheless, the question remains: "How does one go about choosing the right approach for the right situation?"

Just as with choosing the right tool from a tool box, choosing the right improvement tool depends on the full understanding of each tool, how to use it and the situation requiring the improvement. Understanding the different tools is simple. A hammer drives nails. One may also use a hammer to move, compress, smash and reshape. Knowing this does not guarantee that one can probably use a hammer properly, but knowing this might prevent misuse. A knife cuts and shapes. A knife does not substitute for a hammer any better than a hammer will substitute for a knife. Understanding a tool is a start. Being able to use the tool is also critical. Maximum effective use requires experience using the tool in a variety of suitable situations.

Repair is as simple as it sounds. If something is broken, repair it. If a process has a problem, isolate the problem, make a quick fix, then solve the problem. If an external customer receives a product with damage or a defect, it must be fixed. Following this by a study of why the problem occurred and what will keep it from happening again is critical. If the process output is a service and this service does not satisfy a customer, correction is necessary. As with goods, follow correction with the question "Why did this happen and how are we going to keep it from happening again?"

Refine is an approach to making a good product, system, process or activity even better. Refining not only improves efficiency, getting more output with a given input, but also improves effectiveness, producing output which better satisfies the customer with less waste. Many refinements will produce impressive results. Products require constant improvement, sometimes merely to keep pace with the environment, customer expectations and competition. Nothing in an organization is immune to the need for constant improvement and should become an integral part of every employee's daily job.

Renovation is the approach taken to gain a huge improvement. Although the resulting product, system, process or activity might often appear to be much different from the original, it is fundamentally the same. Renovation is a strategy that frequently involves the application of technology, but other approaches are available. Computers, automation, bar coding, new materials, the Internet, overnight delivery, and bioengineering provide some possibilities. A critical aspect of renovation is that the result is transformation, not complete replacement.

Reinvention appears similar to the refurbish strategy, but begins with the premise that the other strategies cannot improve the current product, system, process or activity to meet customer expectations completely. Reinvention is necessary when the aspect of the organization under study is impeding organizational vitality, growth or competitiveness. The initial action in reinvention is to imagine that the current product, system, process, or activity does not exist. Then, armed with intimate knowledge of what the customer requires, the goal becomes to invent a new product, system, process or activity. That which results may or may not have any resemblance to the original, but it is considerable more effective, delivering much higher levels of customer satisfaction.
An example of the four strategies is found in communicating with the printed word. Early books were lettered by hand. Few books were available and then only to the very wealthy. If the printer made a mistake on a page, the page had to be discarded or fixed. Learning how to print better and faster helped. However, books for the masses could never be a reality unless considerable technology could be applied to reproducing books.

The printing press was the answer. Unfortunately, the early printing presses were cumbersome and slow. Many problems plagued the system. As the printing industry solved most of the problems, the modern press began to evolve. Newer presses were fast, could print full-color, and were much more efficient. Still, when a press broke down, it had to be fixed. Manufacturers of printing presses identified the causes behind common failures, designing improvements to eliminate these failure sources. We have seen enormous improvement in printing with the application of new technology. Modern printing is now highly effective, but still does not completely fill the needs of many people wanting immediate access to the information on a printed page. Reinventing the way we get "printed" information to people was necessary. First, the people reinventing "printed information" forgot newspapers, books and magazines. Now, what do many people really want? The right information, despite its nature, where they want it, when they want it. Must it be on paper or a similar material? If not, then do not print it. If the answer is yes, then printing should be an option for the customer. Recently, the information industry as reinvented "printed" material by using electronic retrieval. More recently, the Internet and computer information services have improved information retrieval, dramatically. Keeping the information on printed pages is an option, only if required.

Remember two critical points. First, do not attempt to reinvent most of the organizaions's products, systems, processes and activities at the same time. The other strategies are much more appropriate. Second, shortly after completion of a reinvention project, problems will begin to appear and the need for further improvement will become apparent. As technology changes, the reinvention project may even become a candidate for a refurbishing project, and then eventually, even a reinvention project. Improvement is a never-ending road.

Repair

That which is broken requires repair to function as designed. If the break causes the output of the system or process to be defective, these defects must be fixed. Always use the word "fix" to mean "temporary." Fixes hold things together until an individual or team can identify the causes of the problems, and then develop and implement a solution. Though fixes are short-term measures applied to shore up a process, they do have the habit of becoming permanent.

The critical issue with Repair is the identification and elimination of the root cause. Eliminating root causes insures that the problem, and defective output, will not recur. Repair does not make the system or process better than the original design. For example, an automobile with a defective engine might use a quart of oil every 100 miles. The fix is to add oil when the oil becomes a quart low. This simple fix has not corrected the problem, and will become expensive in time. This leads to a natural problem with fixes. They often become institutionalized. The owner of the automobile might become programmed to stop every hundred miles to add oil and forget that the oil consumption problem still exists. The solution is to repair the engine and correct the defect causing the high oil usage. After rebuilding an engine to meet factory specifications, it is no better than originally designed, unless the rebuilder takes steps to add improvements. All activities to improve the engine beyond the original design are Refinement activities.

Refine

Refining a product, system, process or activity assumes that no problems exist, or that elimination of the problems will take place during the refinement process. Refinement most often comes as many minor improvements in many areas, over a longer period. Refinement is a continual process, and can be the result of individual or team efforts. The refinement strategy involves many individuals in making many small improvements to their jobs. Though creative or innovative thinking will produce impressive improvements, refinement depends on ways to do things just a bit quicker, better, easier, rearranged a bit or the elimination of a small amount of waste. Refinement does not require big "hits" as long as there are many small "hits." Big hits are best left to the renovation and reinvention strategies.

During refinement, products, systems, processes and activities change so gradually that it may appear that no change has taken place. It is much like watching children growing up. The parents do not sense the change, but the grandparents, seeing the grandchildren occasionally, immediately note the change. The beauty of refinement as an improvement strategy is the lack of immediate, comprehensive change. Virtually no resistance to gradual change is usual. However, the managers of the organization may not give the refinement strategy the reward and recognition deserved since the change is so gradual. Another problem with refinement is the tendency not to document minor changes or inform those affected. This effect requires a policy to document and communicate the change to all affected by the change.

Renovate

The renovation strategy provides the real breakthroughs without completely changing the product, system, process or activity. An excellent example of the renovation of a product is the automobile tire. Blow outs and other failures used to plague automobile tires. The tires had little immunity to road hazards. The engineers gradually improved tires, but it was the application of steel belts that made tires dramatically better. Renovation applies advancements in technology and materials to products, systems, processes and activities.

That which is renovated ends up much more improved than had refinement taken place. Refurbishing provides a transformation, a breakthrough. Figure 7-1 shows how refurbishing a product, system, process or activity contributes a higher level of improvement, more quickly. Renovation, however, usually is much more costly than an accumulation of refinement improvements. For this reason, there should not be many renovation projects taking place at the same time.

Reinvention

Sometimes a system, process or activity simply cannot be improved enough to produce an output which meets current customer expectations, let alone exceed current and future customer expectations. Some systems or processes, improved to their maximum levels of effectiveness, still produce a financial drain to the organization greater than the customer benefit derived and revenue provided. These systems and processes must be rethought.

Reinvention projects are almost exclusively assigned to teams. The team begins with vast amounts of customer expectations and a "clean sheet of paper." The new system or process (and perhaps activities) will be constructed using the latest technology and approaches. The new approach will not merely meet current customer expectations at high levels of effectiveness, the new approach will exceed current and potential customer expectations. Though it is possible that reinvention might result in a system or process somewhat similar to the old one, this is generally not the case. This would be the case only if the team chose the reinvention approach when the renovation approach was more appropriate, or the team could not forget the old way of accomplishing the output. In both cases, the team did not succeed, completely.

Figure 7-1, takes a process through repair, refinement, refurbishing, reinvention and back to refinement in an effort to bring output quality to, or greater than, customer requirements. Note how the customer's expectation of acceptable quality kept increasing. This is due to higher levels of customer education or customer expectations being better met by the competition.

Figure 7-1 also shows the concept of "raising the bar" through renovation and refinement. Using the repair strategy to eliminate problems generally brings the output to a point of producing an output which meets the design of the process, not necessarily the requirements of the process customer. Refinement may eventually take the output to a quality level that the customer expectations are exceeded, but much time will be consumed. It is either the refurbishing and reinvention strategy which will take the output beyond customer requirements, fairly quickly. This raises the bar for competition. For public institutions, this creates a stakeholder loyalty and dramatically reduces opposition to the need for increased funding or other changes.

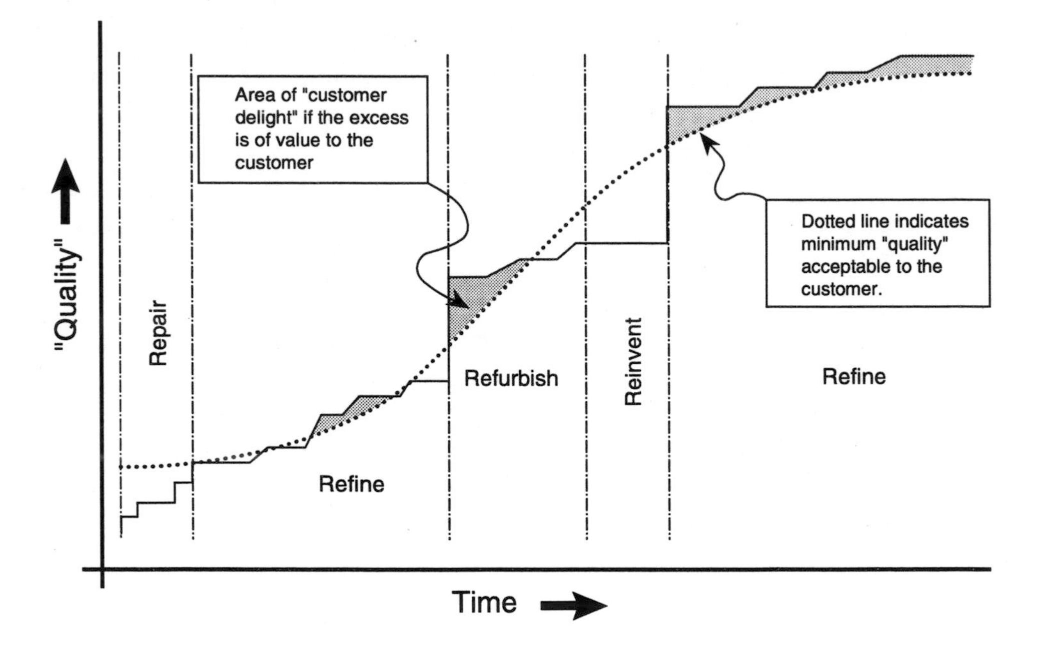

Figure 7-1 Change in improvement approaches over time as the situation requires.

Application of the Four Res

Application of the four improvement strategies involves all people in an organization, from the chief executive officer to part-time, temporary people. This is, of course, within their areas of experience and expertise, and within their authority and span of control.

All processes and activities should be ongoing candidates for repair as problems are identified or are brought to the attention of those with authority over the process. If the problem is major or there is a major effect on customers or one major customer, then a team should be formed to solve the problem. The team must use the seven-step problem-solving process for teams. If an individual is to solve the problem, the use of the seven-step problem-solving approach for individuals is encouraged. Both seven-step approaches are detailed in Section 4. The material in Section 3 will assist the individual or team in making certain the project is the right project and that there is focus. The team will become an effective team more quickly if the approaches in Section 2 are used early in the team's life.

Many problem-solving projects can take place at once, within the limits of available time of the people involved. It is best for the organization or groups within the organization to prioritize the major problem-solving projects for teams to avoid team burnout. Again, Section 3, Project Selection, Direction and Development, will help. All individual employees should be encouraged to isolate and solve problems within their span of control, and implement solutions.

Any system, process or activity with an absence of problems, is a candidate for further improvement. Refinement should be a day-to-day activity. Most refinement efforts should be by individuals. Refinement can be encouraged by a modern suggestion system, with accomplishments recognized by the organization. Some systems or processes, identified as worthy of targeted improvement, could be assigned to a continuous improvement team. However, if the system or process is really worthy of a team's investment in time and resources, the project is better fitted to a renovation project.

Renovation projects are aimed at those systems and processes which need quicker, and greater amounts of improvement. Individual experts can address the innovation needs of the projects. Teams also suit renovation projects. The organization must be aware that renovating a system or process may be expensive. Understanding the cost-payback ratio (Section 3) is critical.

If the organization identifies a system or process which cannot be improved to the extent that it is satisfactory, and may not meet the requirements of the customer now or in the near future, then it is a primary candidate for reinvention. Reinvention tends to require a high investment in time and resources, and therefore, must be a highly informed decision.

Figure 7-2 is a flow chart showing how to integrate the Four Res into the eight-step, problem-solving and process improvement approach. A complete understanding of the current situation takes place in Step Three. At the conclusion of Step Three, the individual or team should decide which of the improvement strategies will be most appropriate to the situation. If the situation is simply a problem requiring a solution, the Repair strategy is best. If the situation calls for some improvement of the current system or process, then Refine. However, the case is often that small to medium improvements simply will not improve the system or process to meet customer requirements, then Renovation is the required approach. Occasionally, a system or process is outdated or defective to the point that no amount of improvement will bring it up to producing an output satisfactory to the customers. Then the system or process must be Reinvented.

Choosing the correct approach is critical. Imagine the high cost of Reinventing a process that simply requires some incremental improvements. Resources would be better used to reinvent a process that is critical to the customer or the long term success of the organization. The same holds true for Renovating a system or process which should be Refined, or Refining something that should be Renovated. This places a high importance on the full understanding of the current situation and what the customers require, now and in the future.

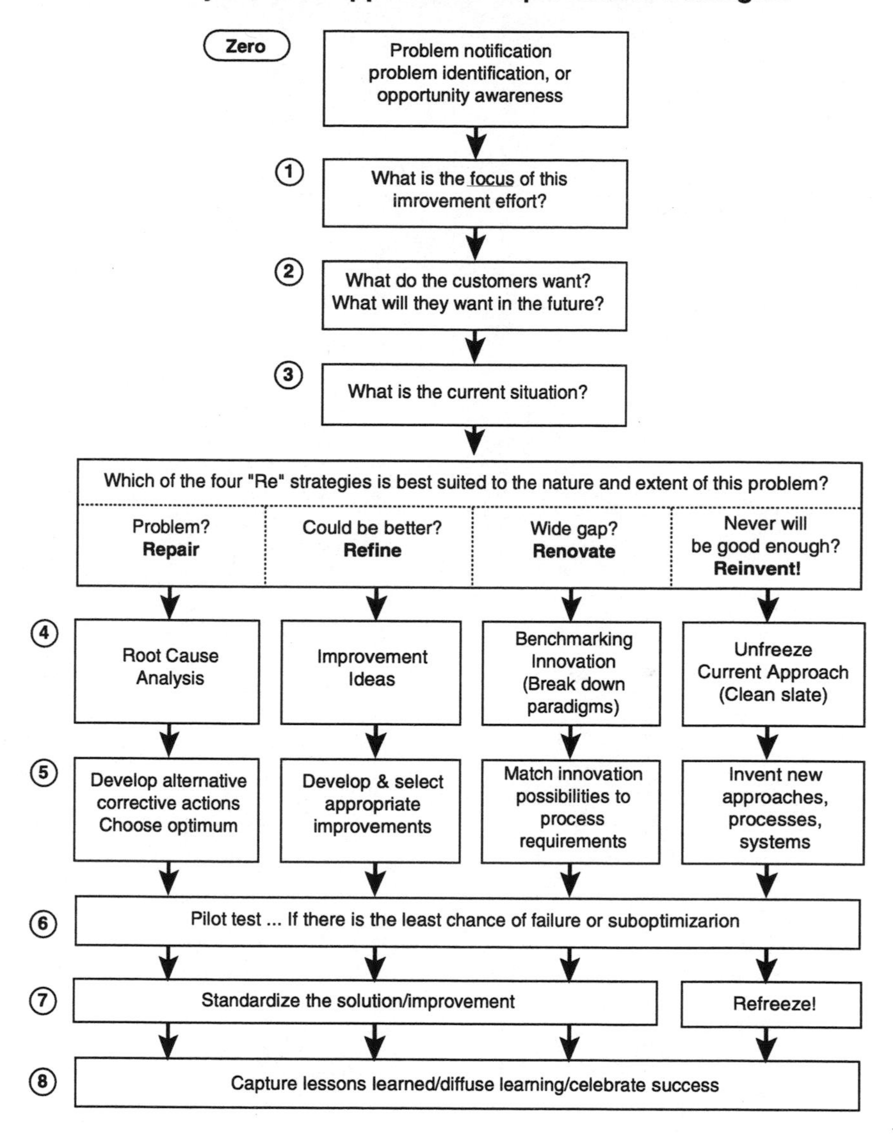

Figure 7-2 Selection flowchart for integrating the Four Res into the eight-step problem-solving/process improvement approach.

Aspect	REPAIR Problem Solving	REFINE Continuous Imrovement Incremental Imrovement (Kaizen)	RENOVATE Innovation Transformation Breakthroughs	REINVENT Reengineering Reinvention Quamtum Leaps
Spark	Results not agreeing with expectations; complaints	Conventional know-how & state-of-the-art applications	Technological breakthroughs, new inventions, new theories imposed on current processes.	Panic: high external pressure from customers, competition & stakeholders: survivalistic
Driver	Meet the customer's requirements and reduce unnecessary costs due to problems and waste	Begin to exceed the customer requirements and improve the efficiency of a process Reduce non-value added costs	Consistently delight the customer, setting new standards "Raise the bar"	Prepare to delight the customers in the future, leapfrogging the competition
Philosophy	*"it's broken, let's fix it "*	*"It isn't broken, let's continue to improve it."*	*"We're doing fine, but let's go for a home run."*	*"Let's completely rethink this."*
Goal	Elimination of the problem, forever	Constant, incremental improvement ... a great many small improvements	Innovation (applying technology to current processes), simplification	Major or total redesign to move into the future
Mode	Correct and maintain	Make better through improvement	Rebuild, removing what is no longer appropriate.	Scrap and build anew
Strategy	Seek out root causes and eliminate them	Find ways to do things better	Apply creative, innovative thinking	Start with a blank sheet and the customer requirements
Focus	Components of a process and individual processes	Components of a process and individual processes	Components of a process individual processes and macro processes	Macro processes and core processes
Mechanics	• Problem identification systems • Idea systems • Problem-solving teams • Individual problem-solving	• Idea systems • Process improvement teams • Process operator/manager initiatives • Experts: individuals and teams	• Idea systems • Problem-solving teams • Process operator/manager initiatives • Experts: individuals and teams	• Reinvention/reengineering teams • Managed project teams
Investment	Generally small for each problem solved	Small to medium	Small to major	Usually major
Payback Amount	Varies considerably	Varies considerably	Medium to large	Large to very large
Payback Delay	Almost none	Little	Some to fair amount of delay	Immediate payback to longer wait
Pace	Slower, though solutions to major problems may be fast-paced	Small steps, often very small, almost imperceptible	Large steps, readily apparent	Huge steps, sometimes traumatic

Table 7-1 The Four "Res" compared.

Noticeable Effect	Solutions to major problems are celebrated, minor solutions go unnoticed	Long-term & long lasting, but undramatic ... can go unnoticed	Short-term & more dramatic, but effort needed to retain gains	Very short-term & quite dramatic: old way is usually completely scrapped
Time Frame	Continuous, as problems are identified	Continuous & incremental	Intermittent, non-incremental	Occasional
Change	Often and intermittent; little change is required except to standardize the solution and educate	Gradual & constant, often is completely transparent	Sudden & irregular, quite noticeable	Abrupt & volatile Dramatic!
Retention of the "old"	Most of old is retained, unless innovation or reengineering is part of solution	Much, if not all, old is retained	Some or most, often layered on to current systems and approaches	None, to little
Effort Orientation	Mostly people	Mostly people	Mostly technology	People & technology
Employee Involvement	Everybody! All individuals, all work groups & many teams	Everybody! All individuals, all work groups & many teams	Select few; several to many teams; includes experts and process owners	Selected few; one or very few teams
Customer Input (Internal & External	Required, since problem is probably impacting the customer	Sometimes helpful, if strategy is to better meet customer requirements	Helpful, if plan is to exceed customer expectations	Critical, especially to identify future customer expectations
Benchmarking	May be helpful	Possible, but not required	Often helpful	Required
Practical Requirements For Success	Must seek out and eliminate root causes of problems, using a wide variety of resources	Requires little investment, but great effort to maintain improvements, since most are small	Requires larger investment, but less effort to maintain	May require very high investement in resources & considerable education
Evaluation Criteria (Measures)	Reduced defects and waste, increased productivity and customer satisfaction	Reduced non-value added cost, and improved productivity and customer satisfaction	Profit & utilization of resources	Customer satisfaction, market share, use of dwindling resources, profit
Disadvantages	Problems often camouflaged. Root causes not identified, thus perpetuating problem even though it appears be solved	Difficult to get excited and commit time, since many small improvements are involved	Could become "technology looking for a use" .. major change is often resisted	Difficult to break with tradition, easy to discard good with bad, could be approach looking for a use
Advantages	Suited to any type growth and economy, much less treatening, "big bang for the buck"	Suited to situations where survival is not the critical issue; appropriate when resources lowest	Better suited to fast-growth economy, but careful implementation and education may then be down-played	Better suited to slow-growth or down-turned economy; erases old paradigms; produces quantum improvement

Table 7-1 continued, The Four "Res" compared.

Section III
Teams

Chapter 8

Team Optimization

Teams

Two heads are better than one. Three heads are better than two and so on. All organizations benefit from the power of teams. However, teams just do not simply happen and management cannot order they perform. Highly effective teams contain the right number of the right people headed in the right direction. They must learn how to work together and overcome the individual tendency to star. They must break down the barriers erected by diversity within the team. They must learn how to use improvement tools and techniques. They must have the resources and authority to succeed. The results will be organizational improvement far beyond what conventional wisdom suggests.

However, take caution. Teams must not take on all problems. Some problems require immediate correction and teams develop solutions slowly. Some problems require the attention of an expert or person of high authority. In addition, overuse of teams takes individuals away from their work, and it is this individual work that provides a measure of individual accomplishment for people.

Given the power of teams and the caution against misuse, use this section to optimize teams.

Forming an Improvement Team

Forming a team right at the start is critically important. Give special attention to the number of members, what they bring to the team, resources they control, and to what extent they want to be involved with the team and the project. Consider seven critical factors when forming a team.

1. Number of members

Five to seven members is ideal. Three or less will probably not include all the resources needed and will probably fail when the group needs creativity. Nine or more may allow the team to become hierarchical. Larger teams permit absenteeism. ("They'll get along fine without me today. I'm pretty busy right now.") Open and equal discussion and presentation of ideas, and concern about team success rather than personal agendas by all members is rare in larger teams.

2. Diversity

Diversity among the team members improves performance, if the team can work through the "storming" and "norming" stages. (See "Stages of Team Growth" next). Variation among the members includes levels of authority, age, education, expertise, nature of their job, sex, race, national origin, and personality traits. Diversity produces creativity by bringing a variety of ideas, concerns, views, experiences, skills and thinking to the meetings.

3. Process owners

If the team is addressing individual process improvement, include a process operator representing that process. In situations where more than one person is involved in the process, choose a person with a better understanding and one respected by the other people; or, let the team choose the representative.

4. Expertise and knowledge

Include one or two members with experience, knowledge or expertise appropriate to the team's mission. Possibilities include the supervisor, a process expert such as an engineer or designer, or a highly experienced, former process operator. The expertise could be found external to the organization. If team membership seems inappropriate to the expert, consider asking the person to become a team resource person, available to the team on an as-needed basis.

5. Resource allocators

Having a member who can garner resources or clear barriers is beneficial. This could be a higher level manager. If this person might bias the team toward the person's preconceived notion of a solution, then ask this person to sponsor the team. The sponsor has the "company checkbook" and meets with the team only by request of the team.

6. Innocence

Consider including someone ignorant of the process or system under consideration. This allows probing ("dumb") questions. They will often recognize problems completely invisible to the other team members. This person must be something of an extrovert, always feeling free to ask the "dumb" questions.

7. Commitment

All members should want to be on the team and have the complete support of their supervisors. This may mean selling the prospective member on the critical need for their membership, exactly why the team needs them and how they might personally benefit from membership. It is even more critical that the reluctant person receives support from their supervisor and their co-workers. Others may have to "cover" while the member prepares for and attends team meetings.

Sometimes, a Team Is Not Best

A team is best if the project cuts across functional boundaries, requires highly creative thinking, or will result in a solution or improvement which process owners must implement. Some projects do not have these requirements. Often, a project remains within a functional area, requires expert thinking more than highly creative thinking and will result in a solution or improvement that will be implemented by an expert from the functional area. Then, assigning the project to an expert in the area with the requirement that the expert gather information from process operators and customers to the process is best.

Traditional Projects

Some projects fall outside the "continuous improvement" initiative. The traditional projects may encompass major change or the creation of something quite new. Examples are new products, facility expansions, acquisitions, and new marketing initiatives. These projects require the traditional project team lead by a project manager. The project manager has greater accountability for the success of the project team than the leader of an improvement team. These project teams, however, will benefit dramatically from the approaches and tools outlined in this handbook, including team optimization tools.

Stages of Team Growth

Teams typically go through five stages of growth. Team formation is the first stage. Bringing together the wrong people or assigning the wrong project, may prohibit the team in moving through the other stages properly. Storming leads to the team norming, the stage where team members develop trust and openness. This naturally leads the team into the performing stage. Performing teams eventually complete their tasks and dissolve, or evolve into a different team. This brings the team to the last stage, mourning.

Stage 1: FORMING

This stage is more critical than it seems. The right number of the right people, bringing the necessary resources, educated and nurtured properly, then given a mission to which they are dedicated, will produce exceptional results.

Improvement teams usually consist of a group of people who do not normally work together, have diverse backgrounds and bring preconceived ideas with them. Feelings range widely, from excitement to disgust with having been selected for team membership. The team does not really accomplish much during the forming stage and that is perfectly normal.

After forming the team, a facilitator should lead the team though some team building exercises. Then the team should develop a code of ethics or ground rules (discussed later in this chapter), and clearly establish its mission.

Stage 2: STORMING

Storming is a difficult and necessary stage for a team. At this stage the team members begin to "let everything hang out." Member ranks are quite evident. The wide range of feelings seems to block the team from ever moving to the next stage. Storming helps team members understand the other members, their similarities and differences, and uncovers their hidden agendas. Storming clears the air and permits the team to get on with its business.

A skilled facilitator is extremely useful during the storming stage and can help the members work through the hurts suffered in the process. However, teams can become dysfunctional during this early stage and remain in the storming stage. The solution to a dysfunctional team is to bring in a highly skilled facilitator or disband the team.

Caution: Take extreme care to prevent the team from trying to move directly from the forming stage to the performing stage. Rarely will a team move directly from the forming to performing stages and become a highly effective team. The facilitator may even need to surface the storming issues to get them on the table and resolved early in the team's life.

Stage 3: NORMING

This is the stage where the team members begin building trust, openness and commitment to the team and the team's mission. Members are still "testing" each other and as members pass the test, the team moves closer to the performing stage.

During this stage of growth, the members reconcile differences discovered during storming. At this point the personality of the team becomes evident. A common direction and spirit evolve. The team becomes more task-oriented.

In a more diverse group, the diversity begins to become a non-issue during the norming stage. Executives and managers remove their "management hats" and learn from the non-managers. Barriers between functional areas, education levels, seniority and personality disappear. Strong leadership becomes less necessary and may become fluid. Team synergy becomes apparent to the facilitator and outsiders as the members begin to operate as one.

Stage 4: PERFORMING

The first three stages merely prepare the team for the important fourth stage. Performing is the desired stage of a team's life, but cannot occur without the team going through the forming, storming and norming stages. The indicators of a performing team are:

- Very low member absenteeism and tardiness,
- Meetings that start on time and end on time,
- Full participation,
- Strict adherence to agendas,
- Evident progress at each meeting,
- Not allowing setbacks to derail the team,
- Transparent leadership, and
- A facilitator with little to do.

In the performing stage, group processes are such that the team operates as one. The strengths of other members reduce individual member weaknesses. Each member supports the others. Commitment between the members is high. Members look forward to meetings and may even become close friends. Discussions are vigorous, open and healthy. Progress is evident.

Stage 5: MOURNING

Most improvement teams face the inevitable: Break up. To a group with high levels of performance and camaraderie, this stage produces trauma. It would be like breaking up a family. Sometimes the team will continue meeting socially or will find a way to keep the team's purpose from coming to conclusion. A skilled facilitator can help the team recognize that they have completed their mission and must move on.

One option in stage five it to plan spinoff projects to capture and use the learning that has occurred. This gives the team members a sense that "life goes on." Team members can "seed" other improvement teams, sharing their experience and insights. Dissolved teams could even plan reunions.

Important Note: Team leadership and team facilitation are two completely different roles. The two roles are different to the extent that the same person cannot fill both functions. The facilitator deals with team process and learning. A team leader deals with project content and actions. The leader keeps the team focus on the mission and links the team with the organization. Early in the team's life the facilitator takes a more dominant role. During the norming stage the leader becomes more central to team activities and the facilitator, more transparent. During the performing stage the need for a facilitator is absent and the leader may even play a secondary role.

Ground Rules/Code of Conduct

Teams need ground rules or a code of conduct to provide a framework for the way they conduct meetings, and treat each other and others in the organization. They need this document before problem situations arise. Construct the code of conduct during the second meeting. Here are some aspects to consider, along with suggestions:

1. **Meeting length and frequency**
   One to one and one-half hours is best meeting length. Schedule meetings for every other week at first and monthly after two months. Or, if there is a sense of urgency, schedule every other week until implementation of the pilot, then twice monthly. For critical projects, consider weekly meetings.

2. **Meeting schedules**
   Regularly scheduled meetings will improve attendance. Schedule meetings several months in advance to allow members to plan around work and personal activities.

3. **Starting late and running overtime**
   Meetings start and stop at scheduled times, without exception. Or, a maximum of 15 minutes overtime permitted beyond the planned conclusion time. This allows members to plan their daily activities better.

4. **Canceling a meeting**
   Half or more members must be present to hold the meeting. For teams consisting of management and non-management members, at least one management and one non-management member must be in attendance to hold the meeting.

5. **Decisions not requiring a team consensus**
   Decisions relating to team mechanics and other non-project issues could be by simple vote. Non-critical decisions require at least half the members present. For critical decisions, all members must be present.

6. **Decisions requiring a team consensus**
   Any decision relating to the team's project will be by team consensus. All voting for team consensus decisions will be by the multivoting process. (Multivoting is presented in Chapteter 12.) Conduct a secret vote if requested by a team member.

7. Systematic approach and team tools

The team will strictly follow the organization's approach to solving problems or improving processes. If an organizational approach does not exist, the team will outline the chosen approach in the ground rules. The use of team tools will not be avoided in the interest of time.

8. Leadership

The leadership team or the person putting together the team could appoint a leader, the team could select a leader by some consensus mechanism, or the leadership could rotate monthly, semi-yearly or whatever, based on the anticipated life span of the team.

9. Recording

The team could elect to rotate recording responsibilities alphabetically or the team leader could ask for a volunteer recorder at each meeting. Since the recorder will have difficulty maintaining a team member role, some teams have found success with using the facilitator as a "flip chart scribe." Of course, this requires the facilitator to be at all meetings and the facilitator may lose sight of facilitation needs. Using an outside person as a recorder is a possibility, also.

10. Confidentiality and reporting

Distribute the meeting report throughout organization by E-Mail, posting, or by another broadcast means. Keeping activities and accomplishments as open as possible is beneficial to the organization and the team. Performing teams could open the meetings to any member of the organization. Visitors would be observers, only. Members may not talk about anyone not present. The team will not permit members getting into personalities. By consensus, the recorder may remove sensitive topics or confidential information from the meeting report for separate maintenance. Organizational policies may determine which topics are confidential, or by a simple majority vote the team could establish that a topic is confidential.

11. Member etiquette and discipline

Members must notify the leader 24 hours (or less, if the team wishes more latitude) before a meeting if they will be unable to attend. Some teams even specify what makes up an excused absence. Any member missing two consecutive meetings without a valid excuse faces expulsion from the team. Assignments must be completed as scheduled. The team must not tolerate raised voices, accusations, swearing, side conversations, interrupting another team member and other counterproductive activities. Continued violation will result in censure or expulsion from the team, as determined by majority vote (a major decision, as noted above). Some teams have successfully used a "fine box." This depends on the culture of the organization and team. Other teams have created a role called "keeper of the code" (similar to a committee's sargent-at-arms).

12. Changes in team membership

Members may elect to quit the team if a valid reason is given. Reasons could be the job or personal pressures, concluding that someone else could bring more to the team, irreconcilable differences, etc. The team must accept the resignation. Some teams do not require a stated reason. The team must decide if "management" will be allowed to change team members and may want the team's sponsor to approve the rule.

13. Ground rule modifications

The ground rules may be modified by majority vote or by team consensus vote.

Meeting Agendas and Reports

Many organizations have discovered that two shorter meetings with good agendas are more productive than one longer meeting. Agendas improve the effectiveness of meetings dramatically. Reports that really cover the essence of the meeting, and the critical details, are read and serve to couple meetings.

Meeting Length

In the early stages of team development, an improvement team will require about one and one-half hours to accomplish what a mature team accomplishes in a one hour. This is normal, and far superior to the traditional three-hour meeting to accomplish an hour's forward movement. Therefore, early in the team's life, the team should schedule meetings for one and one-half hours assuming less effective use of time. As the team begins to function more effectively, the meeting time can be, and should be, reduced to one hour.

Remember that improvement teams generally meet during times that detract from other aspects of the member's job. The correct team makeup, using agendas effectively and succinct reports will improve team effectiveness dramatically. The team's effective use of time and a sense of accomplishment at each meeting will lead to team member satisfaction and few absenteeism problems.

Agendas

Assuming a sixty minute meeting that begins on time and concludes on time, only about 42 minutes remain available for actual activity on the improvement project. The team uses the other 18 minutes for team optimization. This 18-minute time investment more than pays for itself in improved team effectiveness. Here is a suggested agenda for a one-hour meeting:

1. Check-in. (Three minutes)
   - At the first meeting, introductions and possibly a warm-up exercise (for teams where barriers between members must be quickly broken down) should take place. The team members should introduce themselves by describing their role in the organization. Most teams benefit from the absence of rank in the meeting.
   - At subsequent meetings, the meeting should begin with an informal moment for personal or job-related updates, accomplishments, concerns or whatever is on each member's mind at the time.

- Generally, the leader begins or senses that a specific member wants to begin the check-in, and then check-in quickly moves around the table, clockwise.
- This informal moment establishes the readiness to move on.
  (During check-in, each member has a chance to "dump the baggage they bring to the meeting," forming a transition from regular work to team work.)

2. Approve last meeting's report. Allow one minute, since each member should have received the report and read it before the meeting.

3. Review and update action lists for future meetings. What progress is the team making? This should take no more than three minutes.

4. Review and fine tune the agenda. (Three minutes)
   - Additions
   - Deletions
   - Time allotments

These first four points should take no more than 10 minutes

5. Move through agenda items. Watch time!! About 42 minutes will be available for the meeting's agenda items dealing with content.

6. Review accomplishments made at the meeting and the additions to the action list. (Three minutes)

7. Establish next meeting's agenda. Do not forget date, time and place. (Two minutes)

8. Check out: Debrief and evaluate the meeting, noting team process. (Three minutes)

9. Adjourn

Purpose, mission, aim or charter of the team.

Team Name

Meeting Report

Date_______________

Attending	[ ] Person A	[ ] Person B	[ ] Person C
	[ ] Person D	[ ] Person E	[ ] Person F

Agenda Item & Time Allotted
1.
2.
3.
4.

Agenda 1:

Action List:

Agenda 2:

Action List:

Agenda 3:

Action List:

Agenda 4:

Action List:

Next Meeting Date:__________Time:___________ Location:___________

Exhibit 8-1 Useful team meeting report form.

Some team recorders have found the use of the form shown in Exhibit 8-1 more convenient than trying to capture everything that goes on during the meeting. The recorder takes notes carefully, and then immediately after the meeting, makes photo copies and distributes the report immediately. This takes much of the drudgery out of being the recorder and allows the recorder a greater opportunity to be a team member.

Section IV
Creativity Tools for Teams and Individuals

Chapter 9

Generating "Soft" Data and New Ideas

Supplementing or Going Beyond the Data

Often the data does not exist, is limited or exists in a form that is not useful for decision-making. Sometimes the data is distributed among the team members and must be pulled together for the team's use. Next, the data must "speak" to the team members. Finally, data will frequently lead to individual or team vertical thinking, not being able to move thinking to other tracks. This lack of creativity will result in a major block in developing effective solutions or improvements.

This section offers approaches for team members to enhance lateral, creative thinking. Teams and individuals can use either the Five W's and One H or the Cause and Effect Diagram approach to make certain there has been a thorough examination of the situation and that all causes have been considered to the deepest level.

Finally, this chapter presents four techniques that will help teams and individuals overcome vertical thinking. Vertical thinking restricts creative ideas or fresh thinking.

Silent Brainstorming
(An Application of Nominal Group Technique)

Silent Brainstorming is an application of nominal group technique (NGT) and is a highly useful approach for generating many ideas or alternatives while ownership of individual ideas or alternatives remains anonymous. New teams should use Silent Brainstorming during the stage where open interaction is not yet high or when the opinions of one or two people might affect the generation of a variety of ideas. Using the Silent Brainstorming method produces much higher levels of individual team member openness, and therefore, a higher likelihood of fertile ideas or alternatives.

The approach is simple:

1. The team leader asks each team member to take a few moments to think about the issue. This could begin before the meeting by means of a memo. Remember, however, that not all group members will have taken the time to think about the issue before the meeting.
2. Each team member records each idea on a three by five inch note card. Each card must contain one idea, only.
3. The team leader collects all note cards and shuffles them to assure anonymity.
4. The team leader reads each card while the scribe writes the idea on the flip chart. No clarification, lobbying or judgment of any idea is to take place at this point.
5. The team leader asks the team members to generate at least one more card each, pass them in for recording as in Steps Two through Four.
6. The team leader continues to cycle through Steps Two through Five until the members will apparently generate no new ideas.
7. Next, reduce the many ideas to a critical few by:
   A. Clustering (Chapter Five) and then Multivoting (Chapter 12), if there are many ideas (perhaps more than 15) and many seem related in some manner, or
   B. Multivoting (Chapter 12), if there are a fewer than 15 ideas that vary widely.

Brainstorming

If the team is operating in norming or performing stage (See Chapter Eight), Brainstorming is an excellent approach to generating a large quantity of creative ideas or alternatives, but without regard for the quality of the ideas. Many bazaar ideas will naturally release the group's creativity. It is when the creativity of the group increases that new, and extremely valuable, ideas start to surface. During a good Brainstorming session, the team will generate many inappropriate ideas and chaos may replace order. This may be why Brainstorming is not more widely utilized.

Once the team generates a great quantity of ideas, use Clustering (Chapter Five) and Multivoting (Chapter 12) to pare the list down. If, after clustering, the relationship between clusters could provide additional information, construct a Relational Digraph (Chapter Five).

When the team recognizes that many ideas would be appropriate and that the ideas will benefit from creativity, then call Brainstorming rules into effect. A simple approach is for someone to suggest, "Let's Brainstorm it." If the team agrees, the word "Brainstorm" installs new, temporary rules of conduct:

1. Each member of the group offers at least one possible idea, cause of a problem, solution, or suggestion for quality improvement. The team members must keep in mind that it is quantity over quality.
2. The recorder records every idea, preferably on a flip chart, closely duplicating the words of the originator.
3. No value judgments should take place, but members could ask for clarification.
4. The Brainstorming session continues until no new ideas are presented. In essence, everybody "passes." Then, the team leader reads all the ideas and asks the group to read the idea list silently, trying to think about other possibilities.
5. Then the team leader conducts a second Brainstorming session. Highly valuable, creative ideas often come from the second session.

Note: For larger groups (perhaps twelve or more), "round robin" Brainstorming is a better approach. The person to the left of the team leader offers the first idea. Someone records the idea on a flip chart. The leader keeps going around the table, clockwise, until everybody passes. After the group has had a chance to look at and think about all the ideas, the leader conducts a second Brainstorming session, starting half way through the group, and continues until everybody passes again. Often, as in regular Brainstorming, the second round generates highly creative ideas as the team members search for an idea, any idea.

After the Brainstorming session, the "quality" ideas are separated from the others and further developed using the following process:

1. Pare the list is down to the "vital few."
   - Clustering (Chapter Five),
   - Multivoting (Chapter 12).
   - Discussion, then moving on to Clustering or Multivoting, or
   - Subgroup workshop using one of the above three approaches.
2. Then, each of the "vital few" could be brainstormed in terms of causes, solutions, critical components or additional improvement, or . . .
   - Pass on to a problem-solving, process improvement or project team
   - Pareto (Chapter Four) Analysis
   - Etc.

Variations:

Negative Brainstorming: The leader asks for ideas on what could go wrong, what should not be done, what will not work or other more negative approaches. Follow this short session with a positive Brainstorming session.[1]

Challenge Brainstorming: The leader sets the goal of generating an excessive number of ideas in an unrealistically short time. For example: Twenty ideas in two minutes. Two team members may have to do the recording to keep up. To complete the brainstorming challenge in the time allotted, the ideas get wild and creativity increases, dramatically. After a pause, the leader could suggest another minute to collect 10 more ideas, just in case they missed a few.

[1]This approach was introduced by Bill Radford, a quality management consultant from Fort Myers, Florida.

Reverse Fishbone/Improvement Cause and Effect

The Reverse Fishbone allows an individual or team to identify what causes quality. The major difference between the regular Fishbone and the Reverse Fishbone is that the arrowhead on the right points to what the team is desires. The "bones" contain those factors and branches that will cause what they want.

The Reverse Fishbone is useful when planning for improvement. For example, what causes a productive meeting? The Reverse Fishbone uses the same main bone categories that the Cause and Effect Fishbone uses. What materials, methods, facilities, and people issues cause a productive meeting? Environment is a suitable category, also. If the meeting concerns a critical issue for the organization, the Reverse Fishbone might be an excellent meeting planning tool.

Figure 9-1 shows a Reverse Fishbone with the first phase completed. The next step could be to focus on one of the critical first-level causes. The "All employees educated on submission procedures" could produce its own Reverse Fishbone to detect the factors that cause the highest level of understanding of the submission procedures.

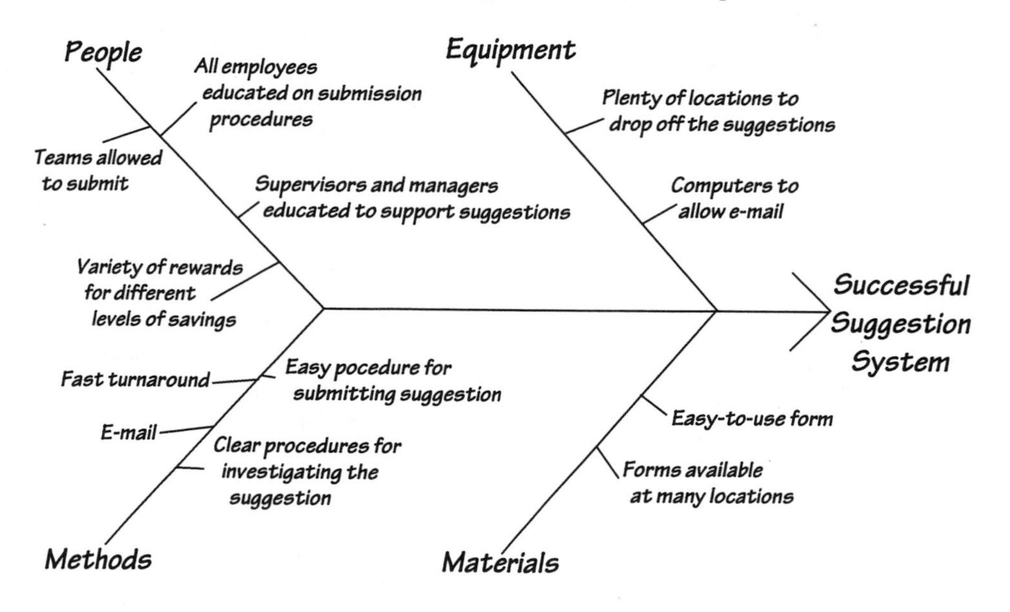

Figure 9-1 Example of a Reverse Fishbone to detect what causes a successful suggestion system

Random Stimulation

One problem with the human mind is that all data and information not considered relevant to the situation or problem tends to be minimized or discarded. This is why Brainstorming sessions often do not result in creative ideas or fresh thinking. It is critical that the team or individual get out of the usual frame of reference. One approach is to use Random Stimulation. Random Stimulation stretches the thinking of the team members way beyond conventional or vertical thinking. Random Stimulation is an excellent tool for individuals since brainstorming requires two or more individuals to stimulate each other.

Random Stimulation is a simple technique that focuses on a word completely unrelated to the issue or problem to help bring about the restructuring of an established pattern of thinking. The mind's attempt to relate this random word to the issue creates new patterns of thinking.

For example, try to relate these two random words: "Dollar" and "Notice." At first they seem totally unrelated. Then, the human mind will find relationships. A dollar is a unit of money and a notice is used to collect money. Or, someone would certainly "notice" "money" laying on the floor.

How does Random Stimulation produce increased individual or team creativity? Normally, if two unrelated inputs are provided, people relate to one and ignore the other as irrelevant. Paying attention to both, as in the above example, forces the individual or team to look for relevance. Just as any idea is captured during a Brainstorming session, all words or ideas generated while searching for relevance are captured. A major difference between Brainstorming and Random Stimulation is that the search for relevance leads the individual or team to fresh thinking rather than a list of alternative ideas.

A Random Stimulation session begins by capturing the issue or problem in as few words as possible. Then, a random word is chosen by selecting a page number and word number. The page number represents the page number of a dictionary. The word number represents a word on the chosen page. For example, the two numbers chosen are 307 and 27. Referring to "Webster's Ninth New Collegiate Dictionary," the twenty-seventh word on page 307 is "crinkle." The issue the team is considering has been captured in the phrase "hazardous waste disposal." At first, there seems to be no relationship between "hazardous waste disposal" and "crinkle." Thinking about "crinkle" in the context of "hazardous waste disposal" will force fresh thinking. The Random Stimulation session could produce:

- Crinkle
- Contract
- Put together
- Confine
- Vacuum pack
- Wrap together
- Smash
- Compress
- Fold many times

Any one of these words could lead to new ways to think about the disposal of hazardous waste.

Interesting enough, the random word can be any word, the more unrelated the better. The object is not to simply build the relationship. The object is to take the individual's or team members' thinking out of the normal "box" within which the human mind normally operates. The longer it takes to build the first few relationships, the better.

Indirect Analogy

Another approach to enhance creative thinking is to attempt to relate two completely unrelated thoughts. As with Random Stimulation, the Indirect Analogy forces creative thinking.

A direct analogy relates a familiar story to a situation or issue. The comparison is obvious in most cases. For example, "The cover of a book is like meeting a person for the first time. What is inside the person is only suggested by the first meeting." Direct analogies are used to better understand something—a concept, situation or condition.

A direct analogy is an excellent tool for learning. A direct analogy is a poor tool for creative thinking. An Indirect Analogy is an excellent tool for producing creative or lateral thinking.

An Indirect Analogy relates two situations that simply are not in any way related. For example, "Bowling is like meeting a person for the first time." One struggles to understand the relationship between meeting someone for the first time and bowling. The analogy is not obvious. Considerable thought is given to each to find a common element. When a connection is made, the analogy become consistent with what we know about analogies.

An Indirect Analogy has some unique features:

1. It has a life of its own. It grows as the individual or team strives to make it a more direct analogy.
2. It is a vehicle for relationships and processes, and these relationships and processes can be generalized to other situations.
3. It provides movement. The problem or situation is carried along with activity associated with making the analogy more direct.

The two requirements of an Indirect Analogy is that it be short and simple, and that the one statement be completely unrelated to the problem or situation.

Consider an organization's problem with dysfunctional teams. The leadership team might be suffering a mental block in an effort to understand why teams have become dysfunctional. Instead of or as a supplement to brainstorming, the team decides to relate "dysfunctional teams" to "grilling hamburgers." Some attempts to make this Indirect Analogy more direct might be:

1. Hamburgers require some fat to be better, but the fat makes the hamburger less nutritional.
2. A team and a hamburger are a mixture of several components.
3. Hamburgers, if not watched, will cause the necessary flame to flare up. Teams also flare up if not watched, or nurtured.
4. Heat is required for a completed hamburger and some heat, in the form of disagreement, might produce a better team.
5. There are several distinct and required steps to cooking a good hamburger. A functional team is the result of several required steps.

After several more analogies are listed the individual or team should sort out what might have been discovered about dysfunctional teams or what produces more functional teams.

An individual or team should follow these steps to use the Indirect Analogy to enhance creative thinking:

1. Chose an analogy completely foreign to the issue or problem.
2. Record ten to twenty more direct analogies. Record every analogy whether it makes sense or not.
3. Look through the list to determine if the issue or problem is better understood. Are there any fertile ideas? Is there anything which might lead to thinking about the issue or problem in a different manner? Is there anything that could be explored further?
4. List the fertile ideas or fresh thinking. Use this list to frame the issue or problem in a new light.

Choice of Entry Point or Shift In Attention Area

All problems command attention and are studied by entering the problem at an obvious or natural entry point.

Consider the problem of a family "totaling" the family's second car. Collecting the insurance for the car, shopping for another one and making the purchase is the natural process for this problem. No creative thinking has taken place. A natural entry point for this problem is how much additional money to add to the insurance payment. Attention is given to purchasing the replacement automobile. Some lateral thinking might take place as the family considers a van or pickup truck.

A much better way to approach the issue "totaled the second car" is to first determine how much it costs to own a second car. This produces a different Choice Of Entry Point. The family can then focus attention on how to best utilize the insurance payment rather than arbitrarily using the money to replace the second car. Another Shift In Attention Area is for the family to consider its life style. Perhaps replacing the second car is the best alternative, but the family will never know if some creative thinking has not be given to other alternatives.

An individual or team should follow this process:

1. Capture the issue or problem in a simple statement.
2. List the complete array of Entry Points. Is there a different Entry Point that makes more sense to study or might produce dramatically different results?
3. Next, capture the issue or problem in completely different terms. One of these Shifts In Attention Area might produce a different way of thinking about the issue.

What is critical about the use of Choice of Entry Point and Shift In Attention Area is that the issue becomes quite different.

Supermarkets often experience long lines at the checkout lanes during certain periods. The natural entry point would be the actual checkout stations. Attention would be focused on the checkout stations. The individual or team studying the problem would probably consider adding more checkout lanes or find ways of speeding up the current ones. Both routes might be expensive and possibly inappropriate.

A different entry point for the checkout problem could be to look at the route the typical customer takes while shopping or what happens after the customer leaves the checkout station. A different attention area might be to focus on how customers get to the check-out lanes or how the supermarket prices its merchandise. Any one of these shifts could lead to some creative thinking by the individual or team.

Remember, the purpose of these creativity enhancing techniques is not to produce answers, directly. The purpose of these techniques it to get minds out of standardized thinking and into alternative ways of thinking about problems.

Section V
Solving Problems

Chapter 10

Problem-Solving and Process Improvement: A Systematic Approach

A Systematic Approach to Problem-Solving

Problems are found in every organization and most organizations want to solve their problems. However, individuals and teams encounter several major hurdles as they strive to solve these problems. First, in the haste to execute a quick fix to reduce or eliminate the effect, overlooking the root cause is easy. Institutionalizing the quick fix is simple, thus erasing the possibility that anyone will discover the root cause. Second, teams and individuals often begin their search for a root cause with a preconceived idea about what the cause might be. This blocks their depth and breadth of thinking, reducing creativity and the likelihood of discovering the root cause. Third, it is human nature to stop at the first level cause and then try to develop a solution. Though better than institutionalizing a quick fix, permitting the root cause to remain could cause the effect or the problem to resurface or produce new problems in other areas. Last, a major problem might require weeks or months of individual or team problem-solving activity. A sense of urgency or lack of progress toward a solution emerges during the problem-solving effort. The frustration results in a solution that is not the optimum solution or one that does not address the root cause.

The systematic approach to problem-solving dramatically reduces these problems and increases the chances of an optimum, cost-effective solution to the problem's root cause.

Problem-Solving and Process Improvement: An 8-Step Approach for Teams

A systematic approach to problem-solving and process improvement is critical to an improvement team's success. The eight-step approach to problem solving and process improvement guides a team through a step-by-step process from problem identification to celebrating problem elimination success. Essentially, the Plan-Do-Study-Act Cycle (PDSA) forms the basis for the systematic approach. (Chapter 14 discusses PDSA.) The PDSA cycle is based on a model for scientific study as popularized by Walter Shewhart, the father of the application of statistics to work processes. The PDSA cycle begins upon recognizing that what should have happened, did not. This initiates a study of the situation. Then, deciding how to act on this information is possible. Developing and executing a plan is the next step, generally. Those charged with implementing the plan, do it. A person or group studies the results to decide how well the results of the implemented plan met or exceeded the expectations established by the plan. The results of this study drives actions. The actions include standardization of what was successful, improvement of what might have missed a bit and correction of what did not work. Each of these actions results in new or modified plans that require execution, then study of the results, and so on until elimination of the problem is the result.

Process improvement initiatives use a modified problem-solving approach. The need for process improvement may not become as evident or as compelling as the need to solve a problem. Customers may not be as ready to provide thoughts about the need for improvement as they are to verbalize problems. The process will not have problem effects to study, nor will there be root causes to identify. The PDSA cycle, however, is equally applicable to improvement activities, as is a systematic, step-by-step approach. The individual or team must keep in mind that the identification of a root cause is not the issue. Identification of improvement possibilities is a critical issue.

The Eight-Step Problem-Solving and Process Improvement approach begins with a Step Zero. This is an additional step to the popular seven-step approach being used in many organizations. Step Zero is critical in that a project must be selected and it must be appropriate for an individual or worthy of a team's time and energies. If the project does not fit the individual or team, or is not important enough, it will not be completed or will result in an inappropriate solution. Often, the project selected is far too extensive or broad to be successfully completed in a single pass. If the project lacks focus, project failure is often the result. Step One is, therefore, even more critical than would first appear. The individual or team must define a project they "can get their arms around."

For example, it is near impossible for a person or team to solve world hunger as a project. However, the right team members or an agricultural expert could solve the problem of growing corn in more arid conditions. Another point to remember is that the project must involve a critical process or critical customer to warrant peoples energies, given the number of problems that exist in most organizations.

Armed with a focused mission, the team should continue Step One by defining what they anticipated or what would have expected. To do this adequately, the team must identify the customers of the process and a determine what they really need or expect. Set measures of success in this step, but only those measures that are important to the customers, or indicate successful improvement or problem elimination. Step One is the "Planning" step because this step establishes the plans for what the project should accomplish. Step One should not be confused with the "Plan" step of PDSA. Step One plans the project concerning what the results should be, not the implementation plan.

Step Two is the "Study" step. Compared with PDSA, it appears that the approach skips the "Do" step. Actually, Step Two studies that which is currently occurring or what will continue to occur if intervention does not take place. Step Two does not study how well planning took place in Step One. A careful focus on the current situation surrounding the project is an absolute necessity. The team must determine the difference between what should be and what is. A great deal of data must be collected. If, during the current situation study, the sense emerges that a quick fix is necessary and possible, as a stopgap measure, the team should implement the fix. The caution here is that the team does not assume that they have solved the problem or improved the process. Fixes are expensive bandages or additional processes layered on the defective process. They are temporary and expensive in the end.

Step Three uses the data and the understanding of the data from Step Two to search out root causes of problems or what gets in the way of highly effective processes. The key word is "root." The individual or team should dig out the root cause with zest. Often, the customer is very perceptive about root causes and can confirm a root cause. Upon discovering what might be a root cause, tested it by asking, "If we eliminate this root cause, will the problem go away forever and reduce waste in the process, while improving customer satisfaction?" Any doubt about a firm "yes" suggests that the team must dig deeper. Since process improvement does not usually involve the identification of root causes, improvement efforts involve determining what is getting in the way of higher process efficiency, effectiveness or customer satisfaction. Where are the bottlenecks, redundancies or unnecessary delays? Can we reduce complexity? Are there opportunities for technology or innovation?

Step Four is much easier if Steps One, Two and Three have been completed properly. Successful identification of root causes usually leads to obvious solutions or improvements. A complete analysis of all factors that limit an otherwise highly effective process will naturally lead to improvement alternatives. Step Four takes the results of Step Three as the basis of developing alternative solutions or improvements. The individual or team must not fall into the trap of trying to implement all alternatives identified in this step. Test one or two of the solutions or improvements, only, despite the tendency to try all viable solutions. The team and process operators can handle only one or two solutions or improvements at a time. Also, if after implementing three or more changes, one or more are defective, it is almost impossible to identify which ones are defective. If the slightest probability exists that the chosen solutions or improvements might not work or could be expensive if widespread implementation were to take place, then a pilot or limited start-up should take place. Step Four concludes with planning a pilot or test implementation, and then executing the pilot or test.

Analysis of the pilot results occurs in Step Five. Hopefully, the pilot was a complete success. If not, what worked and what did not? Why? Are there indications for improvements and corrections? Returning to Step One is common for a team if only to learn if they planned the project properly. Most often, it is because the team took on a project beyond their means. The team might have to revisit Step Two to make certain the team members really understand the current situation. Did they get to the root cause in Step Three or did they stop too soon? If improvement is the issue, did they consider everything getting in the way of optimum effectiveness? In Step Four, were all the alternatives considered? Were appropriate techniques used to sort out the best? Was the pilot or test planned properly? Individuals and teams must not forget that a poor plan, effectively implemented, will not produce improvement. Similarly, an excellent plan, poorly implemented will not produce improvement.

Step Six is the standardizing step. The standardization step implements solutions and improvements full scale, along with communication of the improvement, education and training, documentation and follow up. If the solutions or improvements do not become part of the "fact and fiber" of the organization, people could resort back to the old way as other changes take place. These changes could take the form of new employees, increased activities, downsizing, or changes in the product or service. The improvements might not last due to resistance to change, resorting to the old way because it is easier, lack of understanding, or the lack of leadership attention.

Step Seven is the often forgotten one. After all, Step Six seems to complete the project. However, a great deal of momentum and learning can be lost if Step Seven is not completed. The team has learned much during the previous steps. Capture and diffuse this learning to other teams and people. The root causes and impediments to process optimization discovered during the project may relate to other processes. The solutions and improvements may be appropriate to other processes. Step Seven makes this happen.

Step Seven also provides for continuity. The team or individual may want to take on a new problem or improvement opportunity somehow related to the completed project. The team could be split into two new teams with additional members. The seasoned members could share their learning and experience with the new members, thus producing two new teams with a head start.

Finally, Step Seven provides for the team or individual to celebrate their success. Humans thrive on celebrations. Celebrations reward accomplishments. Celebrations bond people and events together. Celebrations close chapters of life and permit people to move on to new endeavors.

The Eight Step Process for Teams

Step Zero: Project selection/assignment

- First possibility: The organization's leadership team or other higher-level group, or the project team, selects a problem to solve or a process to improve.
- Second possibility: A customer or customers may drop the problem or improvement opportunity onto the organization, team or individual.
- Third possibility: Change, whether externally or internally driven, may call one's attention to the need for solutions to problems or the critical need for improvement.

Note: This is designated as "Step Zero" to keep the approach consistent with the seven-step approach used by many organizations.

Step One: Planning

- Establish the purpose, aim or mission of the project team (whether assigned or individual/team chosen). "What am I (or, are we) really trying to accomplish?"
- Take the necessary time to focus the mission of the team.
- Identify and survey the customers the project will affect. (What's at issue? Who are the customers? What are their needs?)
- Develop an issue statement to attain a critical project focus. Use these four components:
  - — Direction of change or improvement (increase, decrease, reduce, etc.),
  - — Quality Indicator (number of hours/days/weeks, cost, number of errors or defects, complaints, rework, etc.),
  - — Process (What is the activity, process, or system in question?), and
  - — Measure (How much? Reduce to zero, 20%, less than two hours, etc.).
  - — The issue statement is a one-sentence statement of the project aim. Example: Decrease the time it takes for all incoming first class mail to reach the addressee from 24 hours to same morning or afternoon received.
  - — Test: If this one sentence adequately informs others in the organization what the project is all about, and it really captures the nature of the project, then it is a good issue statement. Just imagine someone asking a team member, "What is your team trying to do?" Does the issue statement capture essence of the team's project?

- Give the team a name. The name could be a shortened statement of the mission, a catchy word or two that establishes a sense of the team's purpose, or some abbreviation or acronym that becomes associated with the team's mission. A strong, catchy team name often produces a early sense of being a cohesive team. Beware of what word the initials of the team's name forms.
- Get sign-off of the team's purpose and/or issue statement from the leadership team (steering committee or quality council) or other higher-level entity with a need to be kept informed.

Step Two: Defining the current situation

- Understand the current process.
- Strive to remain within your purpose, aim or mission.
- Select measurement(s) to detect how the process meets customer needs.
- Collect data (Who, where, when, what, how much . . .), but do not forget that data collection limits do exist. Collecting data costs time and money. Get enough data to understand the current situation, but not enough to overwhelm the team. (See Chapters One, Two and Three for data-gathering tools.)
- Clarify the present situation. (Do not fall into the trap of what the ideal situation should be. This comes in Step Four.)
- If a "quick fix" is necessary, now is the time to do it. However, it is only a quick, temporary fix to keep a bad situation from worsening or to keep from losing customer goodwill. The team has not yet determined the root cause, but will in the next step.
- If a simple improvement is obvious and quick to do, now is the time to do it. If these quick improvements satisfy the purpose of the project, the project is completed at this point. This is, however, a rare occurrence.
- Get sign-off of the current situation, especially if the situation could surprise the leadership team or others.

Step Three: Root cause analysis

- Brainstorm all possible causes of the problem (not quick fixes).
  Pare down to the critical few using Clustering or one of the team consensus tools.
- Search out the root causes (going beyond the first level of cause). A cause and effect diagram is an excellent tool but the team could use the "5 Whys" approach.
- See Chapters 11 and 13 for root cause tools and a root cause test.
- Collect data to confirm each root cause, if that data is not currently available.
- Get sign-off of identified root causes, if surprises are possible or the project is critical enough to keep a higher level team informed.

Step Four: Development of possible solutions and pilot

- Brainstorm and consider all possible solutions or improvements.
- Analyze all solutions, isolating the critical few that produce the greatest positive effect.
- Consider the pros and cons of these few. (The Force-Field Analysis presented in Chapter 12 might be helpful.)
- Plan a pilot implementation of chosen solution(s) or improvement(s), a major impact on customers is possible, or one or more team members question the effectiveness of the solution or improvement.
- Consider a limited test if the improvement or solution is more widespread. This will allow more damage control, if the plan or implementation was not optimum. Also, if other changes are taking place in the organization or the external environment is changing, a limited test will allow better control for these variables.
- Get sign-off of pilot from leadership team, and then pilot.

Step Five: Analysis of pilot/test results

- Evaluate the pilot or test results.
- What worked, what did not, and why?
- Make adjustments (and conduct another pilot/test if appropriate), or go back to Step One, Two, Three or Four to learn what was not optimized or what might have gone wrong.
- Get sign-off of how the team will act on the results of the pilot.

Step Six: Standardizing the solution(s) and improvement(s)

- Implement the solution(s) or improvement(s) full scale, if pilot/test was successful.
- Document, communicate, and educate about the changes.
- Plan for ongoing updates and monitoring to make certain the gains stay in place over time and in the face of change.
- Plan for updating and monitoring.
- Get sign-off on what the team accomplished and how monitoring the standardization will take place.

Step Seven: Future planning

- Review the project to learn:
  - — What worked, what did not?
  - — What did the team members learn?
  - — What should have or could have been done differently?
  - — What could this or another team apply to other processes?
  - — What should other individuals and teams know, resulting from this project?
- What happens to the team next?
  - — Move on to similar process, product or problem?
  - — Divide to form two or three new teams?
  - — Take on a new project?
  - — Disband?
- Get sign-off for this project and leadership team buy-in concerning future projects.
- Celebrate success!

The Sign-off Process for Teams

Early in the implementation of organizational improvement strategies, the leadership team or executive team often assigns problem-solving or improvement projects to teams. The leadership team does not have infinite wisdom. However, members of the team represent a higher level in the organization and, collectively, the members do have a better knowledge of the strategic initiatives of the organization, and a more global view of critical processes and issues. This is particularly true if the leadership team has conducted any kind of team activity to identify critical problems, processes or issues. From the list, the leadership team usually assigns projects to teams consisting of people with the resources necessary to solve the problem or improve the process. This project assignment drives Step Zero of the team's problem-solving process.

The fact that the leadership team often assigns a problem-solving or improvement project to a team gives the leadership team more than a casual interest in the team's performance. The leadership team wants to be kept aware of how the project team is progressing to keep improvement activities on track. Higher level people do not appreciate surprises. The sign-off process provides this feedback to the leadership team in a form that is much better than receiving meeting reports. The sign-off process also provides the leadership team the opportunity to recognize the accomplishments of the project team during the project. The sign-off mechanism also allows the project team a formal "reality test" during the project. Teams find it comforting to know that a higher level group understands and recognized what they are doing. Sign-off should not imply that the team requests permission from the leadership team. Sign-off is a communications and recognition tool, not a permission-seeking mechanism.

An organization need not have a formal leadership team to use the sign-off process for project teams. Teams involved in major problem solving or improvement efforts could ask to make brief presentations (updates) to the executive committee or whatever the organization calls the highest level, operational group. For lesser projects, the project team could ask to present its sign-off to a group, department or division manager. Again, the sign-off functions as a reality check, a mechanism to reduce surprises and a means for a higher level authority to recognize the accomplishments of the team during the project.

Step One, the planning step, contains several major activities. After the team members understand the process and what the process customers need, the team drafts a mission or aim of the project, along with an issue statement. This draft is presented to the leadership team by the project team leader to confirm alignment between what the leadership team expects of the project team and what the project teams expects to do. If alignment exists, the leadership team signs off. Otherwise, the project team leader takes the leadership team's reaction and comments to the team for issue statement modification.

Assigning a project with a scope way beyond the capability of the team to handle is common in the early stages of improvement project teams. For example, the leadership team might assign "improve internal communications." This project is a composite of many broad processes. The project team must focus on a component of internal communications and move on to another component after improving the first. Bulletin boards are an example of a focused project appropriate for an inexperienced team. The team could then move on to improving communications downward in the organization as a second project. Focus is critical. Sign off assists the team in focusing and then remaining on track.

At the conclusion of Steps One through Seven, the project team leader reports to the leadership team for sign-off. This process improves communication between the two teams and assists in keeping the project team on course. Thus, when the project team gets to the fifth step, pilot implementation, there will be no surprises for the leadership team members or the project team members.

Also critical is the sign-off for the last step, "Future Planning." The leadership team may want the project team to take on a new project, with or without some member replacement. Better yet, the leadership team may want to "seed" two or three new teams with the experience from the "seasoned" team. This is an excellent organizational learning strategy. Or, the organization may want to disband the team, assigning members to new teams as resources, team leaders or possibly even facilitators, as the need arises.

Eight-Step Problem-Solving Approach for Individuals

The individual problem-solving process is similar to the team approach with several modifications. These modifications relate to some fundamental differences between team and individual "mental" processes, and the size and nature of the problem. Studying a problem from all angles is much easier for a team, considering the many possible causes. This is because a team can easily learn to brainstorm, each member getting caught up in the brainstorming activity, building off other ideas. Members of the team can lobby for their favorite cause or solution, and even play the role of "devil's advocate." When the time decides to select the most likely root cause or optimum solution, the team can use a form of Multivoting to gain a group consensus. Teams consume more resources and, therefore, are more appropriate for larger problems. With the proper mix of team members and the use of team consensus mechanisms, resistance to change is much lower or may be absent.

Individuals, however, focus on problems requiring individual expertise or quick corrective actions. Individuals also work on the many smaller problems; problems not large enough to justify the resources required by a team. Nevertheless, individuals benefit from following a systematic approach. As with the team approach, the individual approach assists in getting an individual focused on the problem, in tune with the customers of the process, digging out the root cause or causes after executing a quick fix, selecting the optimum solution, implementing the solution effectively, and checking periodically to make certain than the solution remains and that the problem has not returned.

The Eight-Step Process for Individuals

Step 0: Symptom awareness:

- Gap between what was expected or is required and what was or is actual, becomes apparent, or
- Problem was assigned to you.

Step 1: Planning for correction:

- What is at issue?
- Who are the customers
- What are you trying to accomplish?
- Draft an issue statement for yourself, similar to the team approach.

Step 2: Current situation/problem description:

- What is normal? As defined by whom?
- What does the customer expect?
- What is the actual situation?
- What is the extent of the problem, the gap?
    - — Is something expected that simply cannot be delivered?
    - — Is it something that was not planned to be delivered?
    - — Had expectations been met in the past and are not now?
    - — Has the customer changed expectations?
    - — Has the environment changed, creating a gap?
- What, exactly, is the discrepancy?
- Is the problem likely to occur again?
- Gather as much data about the problem as possible.
- Remember, there is a point where gathering more data would cause an investment in time and money without a corresponding increase in data useful in solving the problem.
- If a quick fix or interim "solution" is possible, now is the time to implement it.

<u>Warning: You must go on to Step 3. The problem has not been solved!</u>

Step 3: Identify the root cause(s) of the problem:

- Let your imagination run wild concerning all possible causes, and then:
  - — Eliminate those not relating to the problem, and
  - — Cluster those relating to each other.
- Note what is unique about each cluster.
- Also, use process analysis (S•I•P•O•C) to isolate the problem. (See Chapter Three)
- Begin root cause analysis by asking "why?" about the individual problems and clusters at least five times.
- Get input from process operators and/or customers .
- Test the identified root cause(s):
  - — If I eliminate this root cause, the problem will completely go away?
  - — When "Why?" is asked and there is not a deeper answer, it is highly likely that the root cause has been identified.

Step 4: Develop a solution(s):

- Develop a list of all possible solutions; let your imagination run wild.
- Strike out the ones which are not possibilities because:
  - — Too expensive,
  - — Take too much time to implement,
  - — Are much too risky, and/or
  - — Do not relate directly to the problem.
- Cluster the ones which seem to be related.
- Use a Force Field Analysis to isolate the optimum solution(s). (See Chapter 12)
- Important: *If the problem affects a large number of internal or external customers, or is critical to the future of the company, or you have some apprehension about whether the solution will work; then pilot the solution or implement on a limited basis.*
- *Plan a pilot and go to Step 5. Otherwise, go to Step 6.*
- It might also be appropriate at this point to get sign-off from your supervisor or someone high in your organization.

Step 5: Conduct a pilot and study the results (or implement on a limited basis and carefully monitor the affects of the solution):

- As a result of the pilot, what modifications of the solution might be necessary?
- Consider:
  - — Was the current situation not well understood? Go to Step 2.
  - — Was the root cause(s) not identified? Go to Step 3.
  - — Was the solution not optimum? Go to Step 4.
  - — Was the solution poorly implemented? Go to the planning phase of Step 2.

Step 6: Implement the optimum solution(s) full scale:

- Implement the solution(s) on a full scale, permanent basis.
- Who needs to be involved?
- Document, especially since only one person was involved in the project.
- Communicate:
  - — Who needs to know?
  - — What do they need to know?
  - — When do they need to know it?
  - — Do not forget the "why." Use the positive factors of the Force Field Analysis as "sales" points.
- Educate, train!
- This step often involves the inclusion of others. Since the problem has been solved by an individual, others might resist the changes the solution requires, may not give their part in the implementation as high a priority as required or may not under stand the solution. Make certain all those affected by the solution are "tuned in" to the decision.

Step 7: Bring to closure:

- Plan to check back on the solution at future times to make certain the solution is in place and is working. Put "tickler" notes in your planner as reminders.
- What did you learn from this problem-solving effort?
- Does the root cause identification or solutions have application in other areas or with other problems?
- Share learning with others in your organization through a presentation or memo
- Take time to feel good about your accomplishments. Consider a personal celebration! Or, simply reward yourself with something special.

Chapter 11

Digging Out Root Causes

Root Causes

Problems are the result of causes, which are caused by a deeper causes, and deeper causes, until one arrives at the deepest causes, the root causes. Of the root causes, several, if not one, will cause most of the problems.

Root causes are fascinating. If all are eliminated, the problem will simply go away. Also, eliminating the root causes of problems usually provides the lowest cost alternative over the long term. Though executing short-term, quick fixes may be necessary to shore up activities, processes, systems and products of the organization, it is the elimination of root causes which produces the greatest long term pay backs at the lowest cost.

If variation is the issue, one early activity is to detect if the culprit is a common cause of variation or special causes, or both. Since some variation is always present in processes, the search is for the root causes of special causes of variation. Then, the root causes of common variation can be isolated and eliminated. This will produce a predictable, conforming process.

Separating Special and Common Causes in Control and Run Charts

Often an individual or team wishes to learn if special causes or common causes of variation are plaguing a process. Also, the individual or team may need to understand the nature of a process by examining the chart to gain a sense of the variation present.

A simple technique is to stack up all the data points to the left edge of the chart. Figures 11-1 and 11-2 show how this technique clearly indicates the variation in the processes. This variation is not evident in the basic chart. Only by building a frequency distribution will the nature of the variation become apparent.

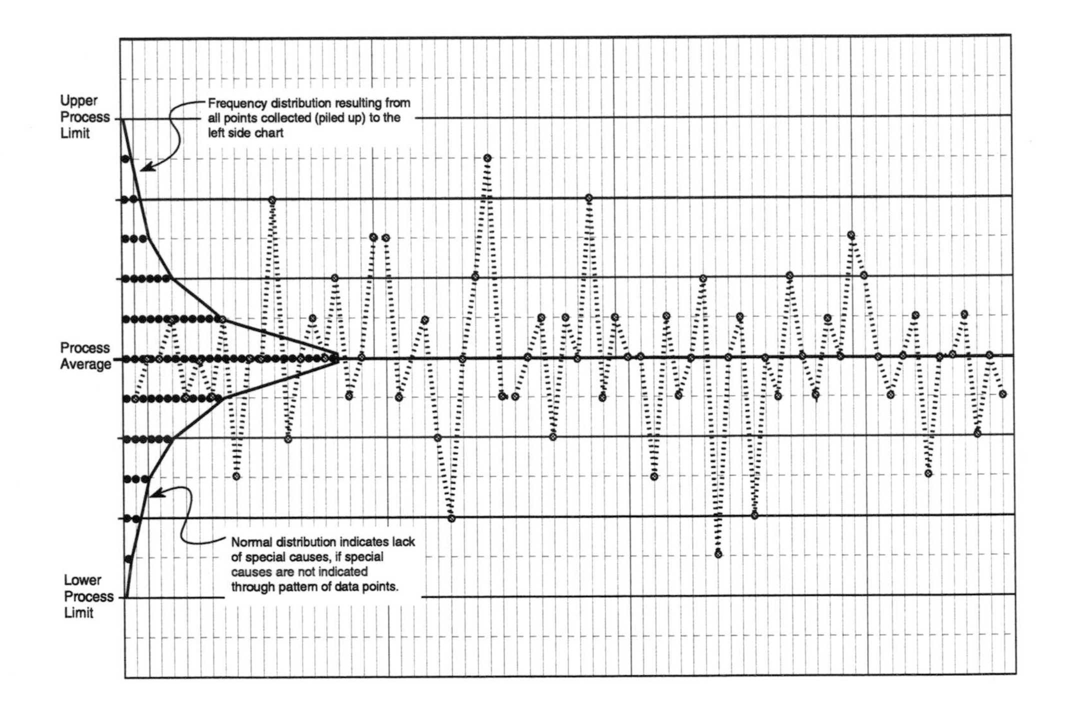

Figure 11-1 Chart with frequency distribution added. The chart suggests normal common variation only.

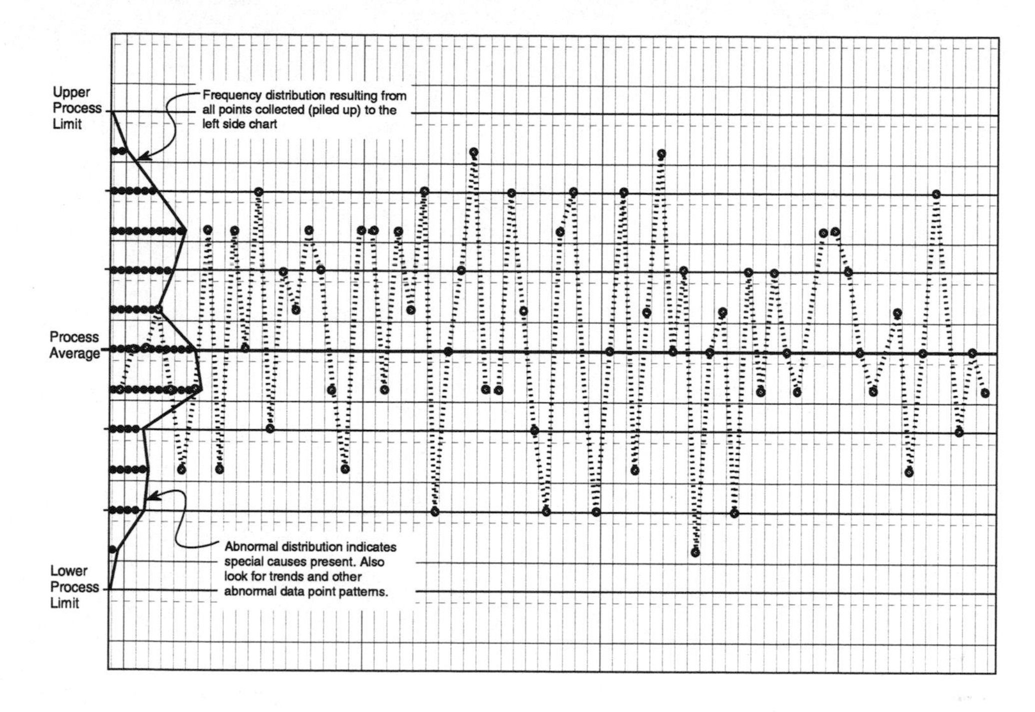

Figure 11-2 Chart indicating abnormal variation.

Chapter Fifteen will discuss Actions on Systems (correcting common variation) and Local Actions (correcting abnormal variation). Chapter Fifteen warns that treating a common cause as special cause or a special cause a common cause will generally make matters worse.

Five "Whys"

One of the quicker ways to dig out a root cause of a problem is a technique practiced by young children. Children often ask "Why?" and when they get the answer, follow with another "Why?" They might even continue asking "Why?" until they get a satisfactory answer, tire of asking why, or hit a obstacle. Unfortunately, most adults get tired of "Whys?" and say "Just because!" or something else to quiet the child. Too often, a child repetitively asking "Why?" is identified as a problem child and dealt with accordingly. This is perhaps why, as adults, we have lost some of our ability to simply ask "why?" until we understand the problem or know the root cause. Also, in our organizational lives we have many activities to perform and take the first answer we get as "the" answer.

Do not be afraid to ask "Why," followed by another "Why," and another "Why?" Why did I cut my finger? Because a sharp knife got too close to it. Why? Because I wasn't watching what I was doing at the time. Why? Because I was in a hurry. Why? Because the job was taking more time than it should. Why? Because I did not understand the extent of the job. Why? Because I didn't take time to ask a couple of key questions before I began to start cutting. I know better now. I must do a better job of planning. Interesting, a couple of key questions about the nature of the repair would have saved my finger.

Generally, if "Why?" is asked at least five times, the root cause of a problem will often pop out. This is the quickest and easiest form of root cause analysis. However, if the problem is complex or the cause well hidden, the individual or team might have to resort to another cause and effect analysis such as the Cause and Effect Diagram.

It must also be realized, there may be more than one root cause to a problem. The "Whys" could branch out like a tree, producing several identified root causes. If this is the case, the root causes must be Paretoized (Chapter Four) to be able to focus on the root cause or several root causes producing the greatest impact.

Here is an example of how the "5 Whys" can be used:

1. Some employees are late for work too often. Why?
   a. Because they oversleep.
   b. Because they are not punished severely enough for lateness.

2. Since punishment is not a solution, or even a quick fix to the problem, the why question must focus on why they oversleep?
   a. They do not get enough sleep some nights.
   b. They do not want to get up and go to work; so they turn off their alarms, roll over and go back to sleep.

3. The team decides that "b" is the cause which should be addressed. Why do they not want to go to work?
   a. They get no satisfaction out of their jobs.
   b. They see their jobs as punishment.

4. "No satisfaction" would seem the most productive route to take at this time, with the team "parking" the "punishment" cause for further analysis if the root cause of "No satisfaction" cannot be found. Why are they not getting satisfaction from their work?
   a. Satisfaction is not designed into most jobs here. We have never considered satisfaction as a component of employment here.
   b. The employees are expecting too much from work.

5. Why have we not built into jobs the mechanisms to allow the employees to get satisfaction?
   a. We do not know what these mechanisms are.

6. Why?
   a. Most supervisors may have the belief that satisfaction need not, possibly should not, be part of what an organization builds into jobs.
   b. We have never considered the link between job satisfaction and performance, including punctuality.

At this point, one root cause of lateness has been identified: Lack of mechanisms to provide job satisfaction. There are other causes of lateness, some of which the organization can address. Note that it becomes apparent that once the root cause is identified, the solutions become somewhat obvious. Supervisors must be made aware that satisfaction is an important part of any person's work and supervisors must also learn what produces job satisfaction and then build some of those mechanisms into the job.

Team use of the "5 Whys"

1. Brainstorm the first "why?"
   Note: Remember the Brainstorming rule that quantity is desired at this point, not quantity.
2. Through Multivoting and discussion, remove those that do not contribute to the "why?"
3. Then cluster the remaining based on commonality.
   Note: If there are many "Whys" listed, or some seem closely related, conduct an Affinity exercise (Chapter Five).
4. Multivote to isolate the critical one or few.
5. Brainstorm the most critical "Why?" for a second level "why?"
6. Continue steps 3, 4 and 5 until all critical root causes are identified.

Cause and Effect Diagrams (Ishikawa, Fishbone)

The Cause and Effect Diagram, conceived by Kaoru Ishikawa, can be applied to any type problem and is highly useful as a team or individual tries to isolate root causes of problems. Sometimes the approach is referred to as the Fishbone Diagram or even the Ishikawa Diagram in honor of the originator. The approach of the Cause and Effect Diagram can be reversed, using the diagram to determine all factors which cause quality rather than those that produce problems. This represents the reverse use of the Cause and Effect Diagram. Chapter 15 presents the Reverse Cause and Effect.

The basic form of a Cause and Effect Diagram begins with four predetermined categories of causes which could be suggested to produce an undesirable effect. These categories are the materials used, the work methods or procedures followed, the facilities, equipment and tools utilized, or the people performing the work. A fifth, and even a sixth factor could be added. Often the group identifies a cause which just does not fit the four basic categories. These causes may relate the to environment in which the process exists or possibly the actual measurements utilized metrics.

The second form of the Cause and Effect Diagram uses categories determined as the result of a simple Clustering session or a more formal Affinity Diagraming session. This approach actually combines the Brainstorm, Affinity, and Cause and Effect Diagram approaches and is referred to as a Generated Category Cause and Effect Diagram.

Basic Cause and Effect Diagram

Begin the basic Cause and Effect Diagram by constructing a bare "fishbone" as show in Figure 11-3. If an individual is attempting the use the Cause and Effect Diagram, the fishbone can be drawn on a sheet of notebook paper. If a team is about to construct a Cause and Effect Diagram, the fishbone is drawn on a piece of flip chart paper or a white board. The flip chart paper allows the diagram to be saved and some electronic white boards produce hard copies of the finished Cause and Effect Diagram. Fill in the four basic categories.

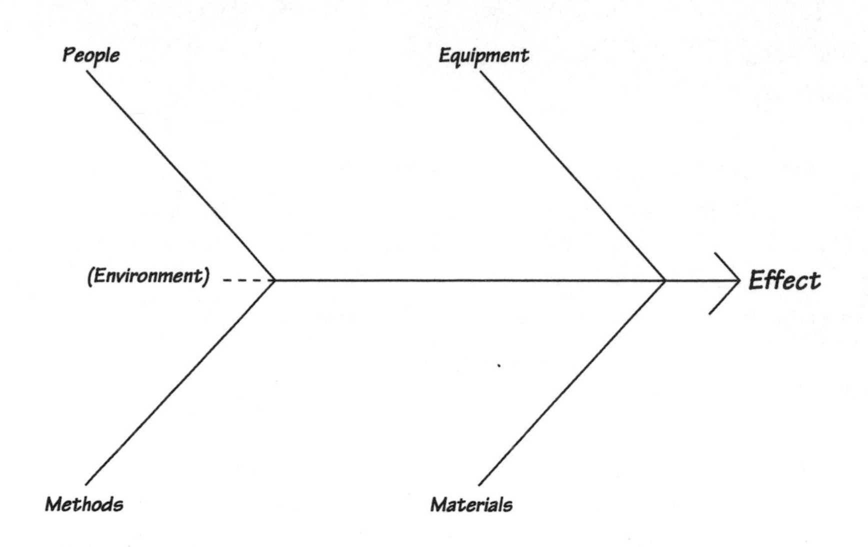

Figure 11-3 The basic Cause and Effect Diagram.

Most often, the cause identified by the first level Cause and Effect Analysis is really an effect which should then be analyzed by a second pass. In the example shown in Figure 11-4, most of the causes for "delivery scheduling problems" could be caused by something else not yet identified or even considered. Applying a solution to a cause which is not a root cause is merely a "fix." Fixes tend to be temporary measures which must be repeated, and add cost and complexity. This is counter to good quality management. When completed, a good Cause and Effect Diagram may appear complicated, but will offer a visual presentation of a vast amount of information.

The above example leads a group to identifying most, if not all, the possible first level causes of late delivery of a good or service. The group would next chose the most likely cause and use another cause and effect diagram to determine all possible causes of that cause. An example of the next pass is shown in Figure 11-5. It is a rule-of-thumb that the root cause can be identified in about five levels of Cause and Effect Analysis. Sometimes it may take only two or three, sometimes six or eight. Note that the second pass of the cause and effect analysis has forced the group to look beyond a preconceived notion of the root cause. This is one of the benefits of the cause and effect analysis. The group might chose to focus on "Scheduling people not aware of who's available and when" as their next first level cause to be analyzed.

The steps in creating a cause and effect diagram are:

1. Decide the quality characteristic relating to the problem or opportunity. What do you want to improve or correct?
2. Write the characteristic on the right and draw a long, horizontal line to it with an arrow on the right end.
3. Write in the main factors or categories which could be causing the problem or limit effectiveness. Group as many as possible. (If more than about six categories result, there has not been enough focus.)
4. Draw in the branch lines.
5. Determine (Brainstorming is an excellent technique) which factors could be causing the problem or limiting effectiveness. General causes will naturally lead to more specific causes. Often, a general factor can be treated as an effect and further diagramed.
6. Use the completed Cause and Effect Diagram as a source of information to solve a problem or improve a process.

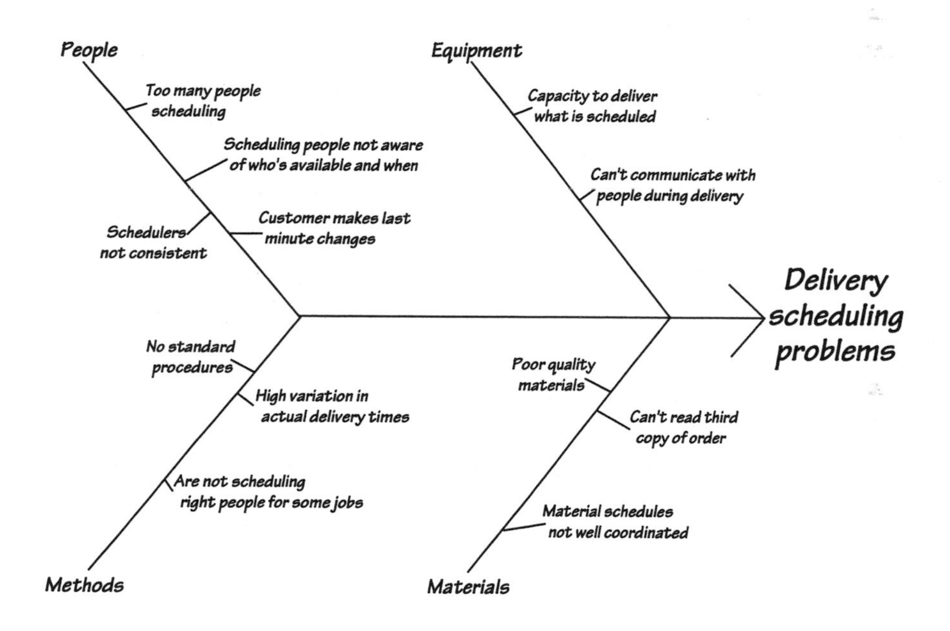

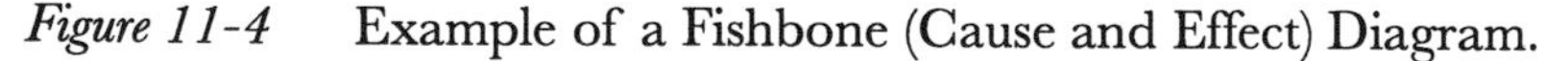
Figure 11-4 Example of a Fishbone (Cause and Effect) Diagram.

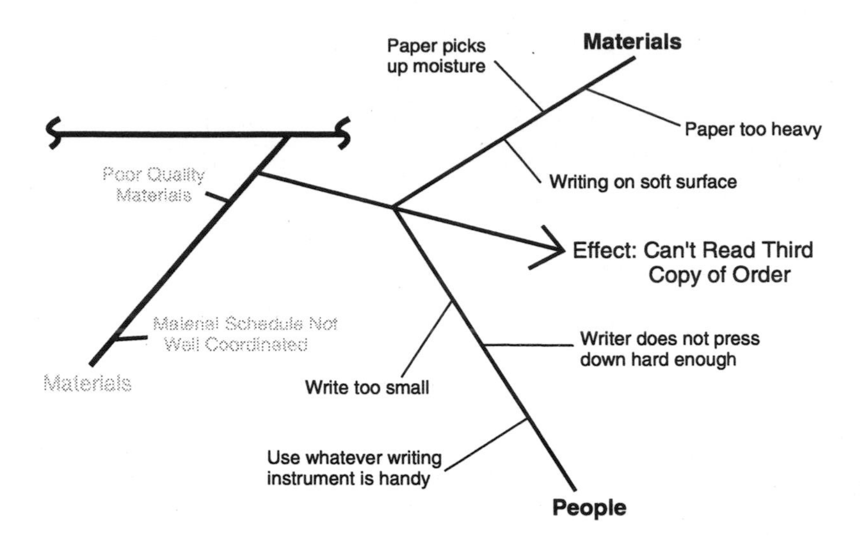

Figure 11-5 Sample of a second level Cause and Effect Analysis.

Generated Category Cause and Effect Diagram

Combining Brainstorming, Affinity Diagraming and Cause and Effect Diagraming produces a Cause and Effect Diagram where the four or more fishbone categories have been generated by the team. This type of Cause and Effect Diagram begins with the selection of an issue. The issue is then Brainstormed to generate a list of possible causes. These causes are Clustered (if there are fewer, clearly understood items) or are subjected to the Affinity approach. The headers then become the main categories of the fishbones. If seven or more categories are produced by the Clustering or Affinity exercise, Multivoting can be used to produce the critical few categories. Once the selected categories are placed on the Cause and Effect Diagram, the exercise proceeds as in the basic Cause and Effect Diagram. An example of the Generated Category Cause and Effect Diagram is shown in Figure 11-6.

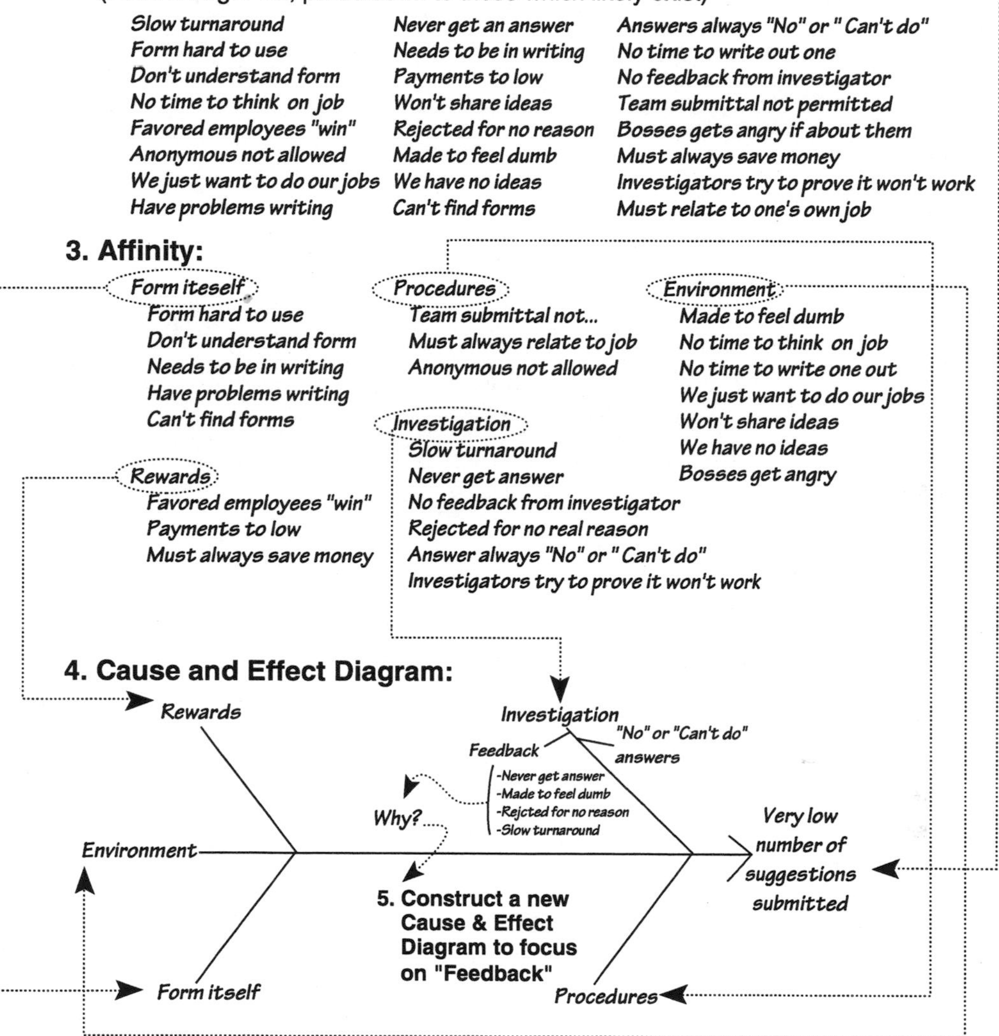

1. Issue statement:
Very low number of suggestions submitted
2. Brainstorm:
(From a larger list, pared down to those which likely exist)
Slow turnaround
Form hard to use
Don't understand form
No time to think on job
Favored employees "win"
Anonymous not allowed
We just want to do our jobs
Have problems writing
Never get an answer
Needs to be in writing
Payments to low
Won't share ideas
Rejected for no reason
Made to feel dumb
We have no ideas
Can't find forms
Answers always "No" or "Can't do"
No time to write out one
No feedback from investigator
Team submittal not permitted
Bosses gets angry if about them
Must always save money
Investigators try to prove it won't work
Must relate to one's own job
3. Affinity:
Form iteself
Form hard to use
Don't understand form
Needs to be in writing
Have problems writing
Can't find forms
Procedures
Team submittal not...
Must always relate to job
Anonymous not allowed
Environment
Made to feel dumb
No time to think on job
No time to write one out
We just want to do our jobs
Won't share ideas
We have no ideas
Bosses get angry
Rewards
Favored employees "win"
Payments to low
Must always save money
Investigation
Slow turnaround
Never get answer
No feedback from investigator
Rejected for no real reason
Answer always "No" or "Can't do"
Investigators try to prove it won't work
4. Cause and Effect Diagram:
Rewards
Investigation
"No" or "Can't do" answers
Feedback
-Never get answer
-Made to feel dumb
-Rejcted for no reason
-Slow turnaround
Why?
Environment
Very low number of suggestions submitted
5. Construct a new Cause & Effect Diagram to focus on "Feedback"
Form itself
Procedures

Figure 11-6 Combining Brainstorming, Affinity and Cause and Effect Diagramming to determine the cause of "Very low number of suggestions submitted."

Reporter's Six
Five W's and an H

News reporters and investigators typically gather information about a situation by asking six probing questions:

1. What happened?
2. When did it happen?
3. Where did it happen?
4. Who was involved?
5. How did it happen?
6. And finally, why did it happen?

News reporters concentrate on the first five questions and bring in the sixth only if it is clear why the situation happened. Investigative reporters and investigators use the answers resulting from the first five questions to support the answer to the sixth question. A good story or investigation results in identifying the culprit who caused the situation. However, unlike good reporting or investigating, good problem cause or improvement investigation identifies the "what went wrong" to highlight improvement opportunities. Focusing on the "who went wrong" is generally counterproductive.

Less critical problem investigations and process improvement initiatives benefit from answers to the same six questions. The structured approach assists a person or team in making certain that the issue is examined from all angles and that enough data is gathered. For more critical problems or improvement initiatives, the Reporter's Six can produce a early understanding of what data must be gathered and where to gather it.

As stated, the Reporter's Six can be used for less critical problems or improvement initiatives. The initial question, before using the Reporter's Six, should be, "Why are we targeting this process or situation for problem-solving or improvement?" This will help establish the amount of time and energy to be allocated to the project.

Then, the initiative can begin with the question, "What is going on here?" Data gathered using the approaches outlined in Chapter Three provides the starting point for the "What" question. Then the team or individual can move on to the when, where, who and possibly, the how. Armed with answers to the first five questions, it is time to move on to the "Why?" Often, it is recognized that the situation or process is not understood well enough to satisfactorily answer the six questions. This is another value of using the Reporter's Six. In process improvement activities, the answer to the sixth question is critical because the answer leads to solutions or improvements. If the "Why?" is not clear, then time must be spent on improving the answers to the first five questions.

The Reporter's Six is particularly useful when seeking the cause of recurring problems.

The questions could be:

1. What, exactly, keeps happening?
   What are the commonalities?
   Is there a common thread among the occurances?
2. When does it usually happen?
   Is there a pattern?
3. Where does the problem usually occur?
4. Who is associated with the problem?
   If more than one person, what do they have in common?
   It must be emphasized that the investigation must focus on what is going wrong, not who is going wrong.
5. How do the problems occur?
   How are the effects of the problem corrected?
6. Finally, why does this problem keep recurring?
   Do not minimize or avoid this amazingly simple approach to digging out root causes. It has and continues to serve reporters and investigators quite well.

Section VI
Optimizing Decisions

Chapter 12

Optimizing Decisions

Decisions

Making an informed decision is not easy. Yet, an informed decision need not be difficult and pays a premium for the time invested. Information is the secret. Information leads to alternatives and the value of each. If one alterative is obvious among many, then the decision is easy. Most often, however, there are several alternatives. Since information may lead to many alternatives, how does a person or team select the best alternative from the possibilities?

This chapter offers four approaches. The Decision Matrix compares a list of alternatives with a list of consequences of each. Not only does the optimum decision become more apparent, but this approach also improves arriving at a team consensus. Generating a Force Field Analysis for each decision alternative is another approach. The Force Field highlights the obvious decision, and the analyses also provides insight into negative factors to be overcome and positive factors to be used in selling or leveraging the decision. Multivoting techniques dramatically improve team decisions. Single voting results in a "winner" not necessarily supported by those voting for the "loser." Multivoting produces a winner supported by all team members. Finally, Return on Improvement Investment (ROII) highlights alternatives which produce the highest financial return over a selected period. ROII provides financial information which is extremely helpful in focusing on the best alternatives and assists in selling the most "monetarily worthwhile" investment of time and money.

Decision Matrix

Use of a Decision Matrix is appropriate when it is necessary to determine the interaction between two or more sets of variables. The Decision Matrix may be also used to indicate the strength of interaction between each variable. The interactions and strengths help individuals and teams make better decisions. For example, one might want to determine how each alternate solution relates to important criteria such as cost, management support, ease of implementation, rapid implementation, etc. The Decision Matrix is an excellent choice for evaluating solution alternatives. A matrix is also useful as a team establishes the degrees of responsibilities for various individuals and groups. This use is not strictly a Decision Matrix, but follows the same format and can be useful in deciding if all aspects of implementation are covered. A sample of this use of a matrix is included later in this chapter and is called a Task Allocation Matrix.

The Decision Matrix format is quite similar to the basic Matrix Diagram Synthesis Matrix presented in Chapter Five. The major differences are its use and the fact that the row and column intersections are quantified so that the rows can be compared. The comparison of each row's total leads to the best option. Figure 12-1 offers a template for the Decision Matrix.

		Dependent					
		1st	2nd	3rd	4th	5th	6th
Independent	1st Factor						
	2nd Factor						
	3rd Factor						
	4th Factor						
	5th Factor						
	6th Factor						

Figure 12-1 The Decision Matrix template.

Figure 12-2 is a Decision Matrix Diagram of what could result if a given choice is made about how to deal with a problem concerning a piece of equipment. The result is the dependent variable because the result depends on the choice made. Therefore, the choice made is the independent variable. Each intersection of a row and column produces a result or conclusion. The result or conclusion could be a simple "yes," "no" or "?" for "I do not know." Or, symbols could be used in each intersection and later transformed to numerical values, or numerical values could be used initially. Figure 12-3 shows the result after the symbols have been converted to numerical values. Then, the importance of each column is weighted to correspond with the desirability of the outcome. This often assists in selecting the best alternative among several good ones.

	Low cost	Impact on process operator	Short-term fix	Long-term solution	Positive customer impact	Feasible within 6 months	Mgmt approval not reqd.
Ignore problem	++	0	0	0	0	++	++
Execute a quick fix, only	++	0	++	0	0	++	++
Quick fix and rebuild later	++	+	++	+	0	++	++
Rebuild now	+	+	+	+	+	+	++
Replace with equivalent	0	+	0	+	+	+	0
Replace with state-of-the-art	0	++	0	++	++	0	0

Figure 12-2 Decision Matrix Diagram using symbols to indicate the nature of each intersection.

	Low cost	Impact on process operator	Short-term fix	Long-term solution	Positive customer impact	Feasible within 6 months	Top mgmt approval not reqd.	Score
Ignore problem	3	0	0	0	0	3	3	9
Execute a quick fix, only	3	0	3	0	0	3	3	13
Quick fix and rebuild later	3	1	3	1	0	3	3	15
Rebuild now	1	1	1	1	1	1	3	9
Replace with equivalent	0	1	0	1	1	1	0	4
Replace with state-of-the-art	0	3	0	3	3	0	0	9

Figure 12-3 Decision Matrix Diagram with symbols converted to scores and a scoring column added to help identify optimum decision.

If the scores do not have at least a three to one range, or some dependent variables are more critical than others, then the dependent variables (columns) should be weighted. A simple approach to weighting is to distribute 20 points among the dependent variables according to which variables have a stronger influence on the decision. In the example shown in Figure 12-4, management has stated that the team must execute the quick fix and/or solution within six months, money is currently available and a positive impact on the customer must be a factor. The selected alternative must reflect these criteria. Figure 12-4 shows how to factor these criteria into a Matrix Diagram. The "Replace with state-of-the-art" alternative suddenly becomes a viable alternative. The intersection "Feasible within six months" and "Replace with state-of-the-art" is the problem. If the team could improve the feasibility of a state-of-the-art replacement within six months, the score for this row would increase and become the obvious alternative. This is where the team must commit some time and effort.

	Low cost 1	Impact on process operator 1	Short-term fix 1	Long-term solution 3	Positive customer impact 7	Feasible within 6 months 6	Top mgmt approval not reqd. 1	Score
Ignore problem	3x1=3	0x1=0	0x1=0	0x3=0	0x7=0	3x6=18	3x1=3	24
Execute a quick fix, only	3x1=3	0x1=0	3x1=3	0x3=0	0x7=0	3x6=18	3x1=3	21
Quick fix and rebuild later	3x1=3	1x1=1	3x1=3	1x3=3	0x7=0	3x6=18	3x1=3	25
Rebuild now	1x1=1	1x1=1	1x1=1	1x3=3	1x7=7	1x6=6	3x1=3	20
Replace with equivalent	0x1=0	1x1=1	0x1=0	1x3=3	1x7=7	1x6=6	0x1=0	17
Replace with state-of-the-art	0x1=0	3x1=3	0x1=0	3x3=9	3x7=21	0x6=0	0x1=0	33

Figure 12-4 Matrix Diagram with dependent variable items weighted to provide more appropriate decision.

The Matrix Diagram has several benefits. First, it helps organize possible choices with potential outcomes so that an individual or team can make an informed decision. When used by a team, the Matrix Diagram also becomes a mechanism for group discussion of the possible choices, the potential outcomes, the relative weight or desirability of each outcome and then arrive at a consensus on each intersection. An added benefit of the Matrix Diagram is that after the optimum item is chosen, the intersection scores for that item leads the team or individual to preventive measures for low scoring intersections and potential selling aspects for high scoring items.

Organizations should use Matrix Diagraming when needing to relate two distinctly different sets of information or issues, and when the choice of an item in one set will affect items in the other set. For example, the choice of where to invest money affects the interest received, how safe the money is, the instant availability of the money and how changes in the economy will affect the growth of the investment. One could construct a matrix to relate the investment alternatives (independent variable) with the result factors just noted (dependent variables).

Matrices can become complicated and difficult to understand. Keep the number of items for each set of variables to a minimum. Six to eight are about the "right" number. The Matrix Decision is useful for both problem-solving and process improvement projects, and can be useful when identifying candidates for renovation and reinvention projects (see Chapter Seven). Remember though: keep it simple! This dramatically improves the daily use by all teams and most individuals.

A common team use of the Decision Matrix to compare a list of project possibilities with the variety of reasons to why each could be a good choice. Another is to compare a list of possible solutions to a list of factors governing or resulting from each solution. The junction of each variable will receive a value or symbol. This allows judgment of each matrix intersection against all the other intersections. If quantitative information is not readily available, such as costs, time, number of customers affected or another measure, or attaining the numerical data would be expensive or time consuming, use symbols. Figure 12-5 offers several schemes for noting the relationship of a column to row.

High	+++	○○	Strong support, strong relationship, excellent possibility, many affected, high cost, etc. (9 points)
Medium	++	○	Some, good, medium. (3 points)
Little	+	△	Weak, possible, fair, a few. (1 point)
None	0	Blank	No, none, zero, instantaneous, forever. (no points)
Negative	-	-	Negative impact or inverse relationship. (-1 points)
High Negative	- -	- -	Extreme negative impact or inverse relationship. (-3 points)

Figure 12-5 Matrix intersection possibilities.

A scientific relationship does not exist among the numbers except that the numerical value of each higher strength symbol should increase by a factor of three to assist in differentiating the stronger intersections from the weaker. If negative consequences are critical, give the single negative three points and the double negative nine points. A team consensus should set numerical values.

Several methods for establishing intersection symbols or point values are possible. Teams could use the "thumbs up" or "fist-of-five" voting process (later in this chapter). Individuals must rely on being prepared with as much expertise and information about the situation as possible and realize that individual decisions are not as effective as group decisions.

Decision Matrix Example

Cnsider the decision-making process used to decide how to handle the dilemma of whether to repair or replace an aging automobile. For example, imagine at 92,000 miles an owner needs to address what should be done with an aging 1976 Ford Mustang Convertible. It is really showing is age (dents, rust, torn top, etc.) and needs considerable mechanical work. What are the alternatives and what factors or outcomes must one consider? The lists of alternatives and outcomes are found in Figure 12-6.

What are the alternatives?	*What must one consider in making the decision?*
Restore the Mustang	*Cost*
Repair to make reliable	*Reliability*
Repair to get by	*Prestige*
Do nothing	*Safety*
Buy used, newer replacement	*Enjoyment*
Buy new replacement	*State-of-the-art, technology*
Buy inexpensive replacement	*Additional features and accessories*

Figure 12-6 The Independent and dependent variable list for the "Mustang showing age" issue.

The set of seven independent variables (alternatives) and dependent variables (considerations) listed in Figure 12-6 form the seven-by-seven L-shaped matrix shown in Figure 12-7.

		Dependents						
		Cost	*Reliability*	*Prestige*	*Safety*	*Enjoyment*	*Technology*	*Features*
I n d e p e n d e n t s	*Buy new Mustang*							
	Restore Mustang							
	Repair Mustang							
	Do nothing							
	Buy a used car							
	Buy an inexpensive new car							
	Buy an nice new car							

Figure 12-7 The seven-by-seven matrix of alternatives for dealing with the decision about what to do with an aging Mustang.

Constructing a two-variable Decision Matrix is not difficult. Follow these steps to arrive at the matrix:

1. Make certain the issue is clear. If the decision is a team effort, make certain all team members understand the issue and the issue is important enough to justify the time required to construct a Decision Matrix.
2. Generate the list of alternatives and possible consequences. This can be done through Brainstorming or Silent Brainstorming offered in Chapter Nine and Multivoting presented later in this chapter. Strive to keep each list at eight or less.
3. Select an intersection scoring scheme and the column (dependent variable) weights. Distributing 20 points among the alternatives is a suggested scheme.
4. Decide what comment, symbol or number should be placed in each intersection. The discussion (teams) or thought (individual) taking place while considering an intersection's score is highly beneficial and provides insight into the variables. Allow adequate time for this to take place.
5. Determine the column weights at this time, if not done earlier. Often the team or individual is better prepared to allot the points after considering all intersections. The one problem with allocating column weights at this point is that it is tempting to allocate weights to force selection of an alternative preconceived as the best course of action before the Decision Matrix exercise began.

6. Total the rows. The row with the greatest point total is the "winner." If the matrix is a team exercise, a discussion should take place at this point to establish that the entire team supports the winning alternative. Healthy discussion at this point is desirable.
7. Concern that the winning alternative might not be the best decision sometimes occurs. One way to address this concern is to refine intersection values or column weights. This must not become an effort to prove that a preconceived idea is the best choice.
8. Use the intersection point values to detect the weaknesses or negative repercussions of the chosen decision and begin preparing contingencies. Consider the larger, positive intersections as selling points for the chosen alternative.
9. The second place alternative is a prime candidate for a backup plan.

Task Allocation Matrix

Often, assigning tasks or allocating resources within a group, department or function is necessary. Specific individuals and groups may be best equipped to handle certain tasks. Also, limited resources place restrictions on who will get what resources. The Task Allocation Matrix approach provides a "picture" of allocations for a project. Figure 12-8 shows how a Task Allocation Matrix improves the allocation of tasks required to implement a plan to improve an organization's hiring process. Note that the matrix establishes the primary and secondary roles for each person or group. The Task Allocation Matrix also points out who has primary responsibility for each task and alerts the team to tasks with two or more with primary responsibility. Each could assume the other is taking the initiative at different times. Additionally, the matrix quickly points out tasks without assigned responsibility.

	Job description training	*Interview skills*	*Job description*	*Job notice*	*Recruiting*	*Screening*	*Selection*	*Orientation*
All supervisors	S	S	P	S			P	P
Director of Human Resources	S			S				
Employment Specialist			S	P	P	P		
Human Resources Assistant					S	S		
Training Specialist	P	P						S

P = Primary role, S = Secondary or support role

Figure 12-8 Task Allocation Matrix for assignment of responsibilities in a hiring improvement initiative.

Multivoting Approaches

Traditionally, most committees use the single-voting approach. Each member votes for their favorite and the item with the largest number of votes wins. If a single alternative gets more than half the votes, assume that it is the unanimous choice. If an item on the list gets more votes than any other item, it still wins. But, do not confuse "unanimous" with "consensus." Not all team members will necessarily support a unanimous decision. Some team members may become "dropouts" for the remainder of the project, or worse yet, may even sabotage the project. Additionally, if the list is large, single-voting breaks down unless there are a few obvious choices. The use of single-voting simply does not work well in a team environment. A much more effective voting approach is to use Multivoting, "Fist of Five" or "Thumbs Up." The weighted voting is also useful for teams desiring to make one or two alternatives obvious.

Multivoting

Multivoting is a useful method to reduce a larger list of possibilities to one team-supported item or a vital few items. The team accomplishes this through one or more voting sessions, with each team member getting two or more votes. The result is a "winner" supported by all members, even if some team members support the winner on a conditional basis.

After generating a list of possibilities through Brainstorming or Silent Brainstorming, and grouping like items, each team member gets two or more votes, based on the number of items on the list. For example, each member could get votes equalling one-third the total number of items on the list. A 30-item list suggests 10 votes for each member. Since Brainstorming or Silent Brainstorming usually results in a large list, an alternate method is to allot the number of votes according to table shown in Figure 12-9.

Items on List	Number of Votes
3 to 6	2
7 to 9	3
10 to15	4
16 to 20	5
21 to 29	6
30 or more items:	7

Figure 12-9 Number of votes to allocate to each team member depending on the number of items or alternatives.

Using the table will reduce the list faster, possibly with a single Multivoting session. If the list is too large after the first vote, conduct a second vote and even a third vote to surface a team choice.

Several Multivoting mechanisms are available:
1. The team leader could simply call for a show of hands for each item, or
2. The members could go to the flip chart and place hash marks next to their choices, or
3. Each team member could get an appropriate number of colored, self-adhesive dots to place next to their choices, or
4. Each member could write each of their choices on a separate note card, passing the cards to the team leader or vote counter.

The choice of voting methods depends on which would be more efficient and whether or not the members want anonymity. An important point to remember is that a team member can give only one vote to each choice. Otherwise, the voting becomes weighted or degenerates to single voting or weighted voting.

Weighted Multivoting

Sometimes the team may want to weigh Multivoting to emphasize the winner. This variation permits the team members to give any item two or more of their votes. Occasionally, the winning item may get so many votes that it becomes a clear winner. If a member of the team realizes that a few of the items have widespread support and the team members may have to cast some votes for items they cannot simply support, Weighted Multivoting may be the best choice. The problem with Weighted Multivoting is that if all team members allocate all their votes to one item, the voting then becomes single voting with all the disadvantages of single voting. For this single reason, avoid Weighted Multivoting.

If Weighted Multivoting might be the optimum approach, the hash mark, adhesive dot or note card method of voting are alternatives.

Fist of Five:

The Fist of Five form of Multivoting will pare a shorter list, quickly. Each member votes on their commitment or feelings about an item by holding up their hand showing all five fingers (strong) to a fist-with-no-fingers (showing very weak or no support). The leader merely counts the total fingers shown by all team members for each item. Fist of Five is also a shortcut Weighted Multivoting approach and works nicely for a less critical decision.

Thumbs Up:

"Thumbs Up" is a quick approach to voting on any size list with some weighing of the vote. A thumbs up vote indicates a member's stronger commitment to the item and receives a weight of three. A thumb positioned sideways indicates weaker commitment and receives a weight of one. Thumbs sideways could also suggest that the team member will support the alternative or live with the decision. A thumb down position indicates "no" and receives a weight of zero.

Interesting group dynamics occurs during the Thumbs Up voting. A wavering side thumb sends the message that "I could be swayed" or "I'm really uncertain." In fact, the voting can change as team members glance around the table to check how other members are voting, and then modify their vote accordingly. This points out a disadvantage of Thumbs Up: group-think. Group-think is a situation where a member representing a higher level in the organization or a member with much higher respect, sets the tone for the voting. Often, members are not aware that their voting is being influenced by another team member. If group-think is a possibility, a secret form of voting is better.

Force Field Analysis

Two types of Force Field Analysis are useful when attempting to analyze the pros and cons of a decision. The first is useful in analyzing the forces working for and against a single issue. This is the "Opposing Forces" Force Field. At times the relative weights of the pros and cons of each course of action or solution become the issue. The second type of Force Field is then used to isolate the optimum decision and highlight the pros and cons of the selected solution. This is the "Pros and Con" Force Field.

Opposing Forces Force Field

Often, opposing forces work on an issue. A variety of circumstances could act to cause problems or produce barriers. Fortunately, opposite forces help or accelerate. The Opposing Forces Force Field provides a visual presentation of the opposing forces and allows an individual or team to understand what is taking place and judge which forces will prevail.

Construction of the Opposing Forces Force Field begins with an activity that generates a list of barriers and aids. Individuals can collect data from the process or system, or can collect information from the customers of the process or system. Teams can supplement data collected by individuals with a Brainstorming exercise. Then, list barriers, or negative forces, on the left-hand side of a vertical line representing the force field. On the opposite side, list the forces creating aids, or positive forces. Figure 12-10 shows a completed force field associated with a "current situation" analysis, Step Three of a problem-solving initiative.

Barriers
Aids
Customers demand fast response
Sales exceed capacity to deliver
Complaints cause extra workloads and this makes full resolution difficult
Sales people have not establishing full needs of our customers
Competitors experiencing same problems
We are highly responsive to complaints
Our employees want complaints reduced, preferrably eliminated
Cost to correct problems quite high
We really value our customers

Figure 12-10 Opposing Forces surrounding the current situation of increased customer complaints.

Analyze each by looking at those forces driving the decision in a positive direction and those driving the decision in a negative direction. This analysis could start with a Brainstorming session to develop a list of all pros and cons, despite their possibility or weight. After identifying all factors associated with the issue, categorized each as more positive or more negative. List the negatives on the left-hand side of the force field and the positives on the right-hand side[1]. Give a weight to each based on how positive or negative it is, ranging from strong effect equals five points to weak effect equals one point. To make certain that many positive or negative effects given two to four points do not overload the analysis, do not hesitate to give minor effects one point or two points.

Figure 12-11 shows an example of a force field analysis for converting some work groups to flex-time.

Negatives	Points	Points	Positives
More difficult to supervise people	-4	3	*Longer workday for customers to contact us*
Fewer times during day for meetings	-2	5	*Better meets the varying needs of employees*
Everybody wants the same work hours	-3	2	*Better use of expensive equipment*
Replacements may want different hours	-1	4	*Employees have asked for flextime*
Employees may not adapt to longer workdays (4 10-hr days)	-3	4	*More flexible scheduling for varying workloads*
Employees may want to change too often	-3		
Vacations may leave certain hours or days open	-1		
Totals:	-17	+18	

Figure 12-11 Force Field Analysis example.

[1]. Some organizations prefer to list the positives on the left side and the negatives on the right side. This is no problem as long as each team member understands which side is opposing change and which is assisting.

In the Force Field Analysis example (Figure 12-11), there is not a compelling positive force suggesting a change to flex-time. Should the positive score exceed the negative score by twice or more, then implement, or at least pilot, flextime. What the example does point out is that several negative factors should be examined, and reduced or eliminated. A team should start with the "difficulty to schedule" issue to decide if better scheduling is possible. Then the team should move to "employees may want to change too often" to reduce that problem and on to higher negative issues. At the same time, the team should examine "better use of equipment" to see if this issue could become a more positive benefit. After these actions, the Force Field could then be updated to determine if the positive forces outweigh the negative forces.

The force field analysis is a powerful tool to compare two or more alternatives to determine which has the highest ratio of positive to negative. As in the flextime example, if the desired choice is not the "winner," the team can study ways to reduce the negative factors while increasing the positive factors.

After the choice is made, the force field alerts the team and the organization to the negative factors which could spell disaster to the solution or new system. In the flextime example, considerable effort should be made to plan an effective scheduling system and to reduce the possibility that employees will want to change too often or may not adapt to different or longer hours. Up front communications becomes an imperative. The positive factors become excellent points to sell the change and to assist in overcoming the negatives.

Return on Improvement Investment

Most efforts to solve problems and improve processes require an investment. This investment consists of time, money, human effort, risk, and lost opportunities in other areas or initiatives. The possible return includes increased revenue, increased profit, improved utilization of monetary resources, reduced human effort, reduced risk, higher levels of customer or client satisfaction, increased market share, and improved vitality of the organization. Some of these factors can be stated in monetary terms and some cannot. Regardless, increased quality and productivity require investments and result in organizational improvement.

Traditionally, organizations measure success based on financial terms. If a problem-solving or improvement effort is financially analyzed to show that the financial return is greater than the financial investment required, the initiative has a far greater chance of being funded by the organization. Since time is money, members of the organization will commit adequate time if the return is worth the time investment. Therefore, it is prudent in most organizations to compute, or at least estimate, the Return on the Improvement Investment or ROII.

Unfortunately, most organizations do not determine, record or track the monetary figures associated with many of the factors which affect the investment in problem-solving and improvement efforts. For example, time spent in meetings is neither recorded or costed. It is easy enough to compute that five people making twenty dollars an hour, meeting for two hours to solve a problem will cost the organization $200. If the implementation costs $100, the total cost of solution is $300. Now, if the problem was a one-time problem resulting in a $100 expenditure to fix, its solution cost the organization was $200 more than was saved. If the problem had been occurring an average of 20 times a year, the $300 solution will save $2,000 per year, a good investment.

Many of the components included in the ROII can be found in Chapter Six. These are the Cost of Non-Quality or CONQ. These are costs that can be included in the future savings produced by the solution or improvement. The improvement investment costs include, but are not limited to:

- Time spent in meetings, conferences, travel and all other off-the-job time spent associated with the solution or improvement activity.
- Any money spent to achieve the solution or improvement. Examples are materials, samples, pilot studies, printing, equipment used for the project, consulting services.
- Prevention measures.
- Implementation and follow-up costs.

Once the CONQ or investment cost figures are accumulated, the ROII figure can be computed. ROII is simply the CONQ for a future period divided by the ROII. For example, if a problem is allowed to continue, it could cost the organization $125,000 in the next 12 months. Developing and implementing a solution is judged to cost $2,700. The ROII is $125,000 divided by $2,700 or 46.3. For the next year, the reduced cost due to the problem is 46.3 times greater than the cost to develop and implement a solution. That is a first-year return of 4,630 percent on an investment of $2,700. What organization would argue that this is a wise investment, assuming that the time and funds are indeed available?

An alternative approach to ROII is to determine the solution or improvement investment as above, but instead, divide the investment figure by the favorable outcome. For example, if a university invested in a teaching improvement effort costing $25,000 and the result would be the graduation of ten additional people who would not have otherwise graduated, the ROII would be ten divided by $25,000. The ROII for the teaching improvement effort would be one additional graduate for each $2,500 spent on the improvement. The decision to engage in the effort could be driven by the value of each additional graduate. Is an additional graduate worth $2,500? If so, the effort is worth the $25,000 investment.

Chapter 13

Testing Root Causes and Solutions

Tests

Tests are similar to analyses except that they provide feedback to an individual or team concerning achievement to that point. Specific points exist during the problem solving or process improvement process where an individual or team must pause and make certain they are making optimum progress. These two critical junctions in the problem solving process are evident: After identifying the root cause and after developing a solution or solutions. This chapter helps test the results up to these junctions.

Testing Identified "Root Causes"

Teams and individuals must identify root causes as a critical step leading to the development of a solution or solutions that eliminate the effects of the causes. In addition, root causes will naturally lead a person or team to effective solutions. Nevertheless, how can one determine that a root cause has been found? The answer: upon eliminating the cause, the effects will disappear and the solution will be cost-effective. For example, if slowing down will eliminate most mistakes in future reports, "going too fast" was the root cause. If mistakes continue to occur, then going too fast is not the root cause. Why the report writer is going too fast, or other causes of mistakes must be found.

Sometimes the solution to a cause (not a root cause) creates new problems. In the previously mentioned example, if the assumption is that the root cause is "going too fast," and the process is slowed as a solution; new, undesired effects may creep in. The report may be late, incomplete or poorly researched. Identifying the root cause or root causes of mistakes in a report and implementing corrective actions will not build in new problems.

When a team or individual determines what may be the various root causes, each root cause should be subjected to a test. Apply the "Test of Root Cause" to each root cause.

Test of Root Cause

Make certain that each question results in an unqualified "Yes."

1. Will the problem go away, forever, upon elimination of the cause?
2. Will productivity or effectiveness improve upon removal of this cause?
3. Is the solution now more obvious, perhaps even indisputable?
4. Has there been complete identification of the "Who," "Where," "When," "Why," "What," and "How" with a high degree of certainty? (Remember, the search is for "What is going wrong," not "Who is going wrong.")

Any question with the answer "no," "maybe" or "not certain" should lead to a reexamination of the root cause analysis. Ask "Why?" one more time or create a Cause and Effect Diagram for this cause.

Make certain that "No" is the answer to each of these questions:

1. Would the likely solution layer on additional costs or more activity (such as additional inspection)?
2. Would the likely solution make the process or system more complex?
3. Will the solution add delays or longer turnaround times?
4. Can we ask another "Why?" and get further insight into the root cause? Or, can we construct another Cause and Effect Diagram and go to a deeper level?
5. Will new problems creep in upon implementation of a likely solution (or solutions) believing that the root cause has been found?

If each root cause does not pass all these tests, go at least one level deeper. Test the next level solutions until all pass the test.

Test of Optimum Solution

Often, if the root cause is well identified, the solution (or solutions) becomes more obvious. The problem at this point is twofold. First, more than one possible solution may exist. Several factors such as developmental problems, cost, speedy start-up, resistance to change, and short-term and long-term effectiveness may vary considerably among alternatives. Second, the selected solution may have positives and negatives. The positives are what make the solution a good choice and allow easier selling of the solution to management, process operators and customers. Positive aspects are assets or aids. Negative aspects degrade the effectiveness of the solution or may hinder implementation. The negatives are liabilities or barriers.

Several approaches are available that allow comparing several solutions. Two approaches, Decision Matrix and Force Field Analysis were presented in Chapter 12. These are very useful in selecting the optimum solution or solutions. Beyond these approaches for identifying optimum solutions and determining the assets and liabilities of chosen solution, answer these critical questions:

1. Does the solution add complexity to the activity, process, system or product?
2. How does the customer feel about the solution? Will the solution please, perhaps delight, the customer? Or, will the customer perceive the solution as a solution or some type of fix? Worse yet, will the customer perceive the solution not as a solution at all, but some form of propaganda?
3. Is the solution long-term, or temporary, while awaiting a better solution? Temporary solutions can be expensive and often frustrate process operators who want problems solved.
4. Is the chosen solution a substitute for an obviously better solution because of limited time, money, people resources or because of panic? Consider this a stop gap solution.
5. Is it possible that the perception will be "the medicine worse than the disease"?
6. Is it known that the cause is clearly common or special? If common, is the solution related to a system or management change? If special, is the solution directed toward a specific person, process, supply, piece of equipment, customer? Strive for a "yes." Note: Most problem-solving activities should be directed toward common causes. Special causes may happen on such an infrequent basis as not to warrant a major problem-solving effort.
7. Will the activity, process, system or product change in the future, eliminating the need for this solution? If so, perhaps a fix is better. Make certain the solution is easy to implement, has an immediate effect and is cost-effective.

Answers to these questions will give insight about the effectiveness of the solution and whether or not implementation should begin.

Section VII
Improvement Strategies and Tactics

Chapter 14

Learning and Improvement

Learning and Improvement

Learning and improvement are closely intertwined. Most advancements made in our world have come from an individual or group learning. The learning process begins by thinking about something different, planning to test the hypothesis, trying it, observing the results, testing the results against the plan, then acting on the difference between what was planned and what resulted. This is the scientific method and is how humans learn. Improvement follows a similar route.

Plan, Do, Study, Act (PDSA)
Shewhart Cycle, Deming Cycle

In his quality control lectures in Japan in 1951, W. Edwards Deming used a learning cycle based on Walter A. Shewhart's three stage cycle of activities necessary to manage a production process: specify, produce and inspect. These three steps parallel the steps of the scientific method: hypothesize, experiment, test the hypothesis. The plan, do, study, act cycle being based on both, is, therefore, a process for learning and improving.

From Deming's lectures, the plan-do-check-act approach evolved. The Japanese often call the cycle the "Deming Cycle." At other times, people call the cycle the PDCA cycle. Purists often call the cycle the Shewhart Cycle. Originally, people called the third step "Check," but "Study" more accurately describes what happens in Step Three. Many people now call the cycle Plan, Do, Study, Act, or PDSA. Figure 14-1 shows the basic PDSA cycle.

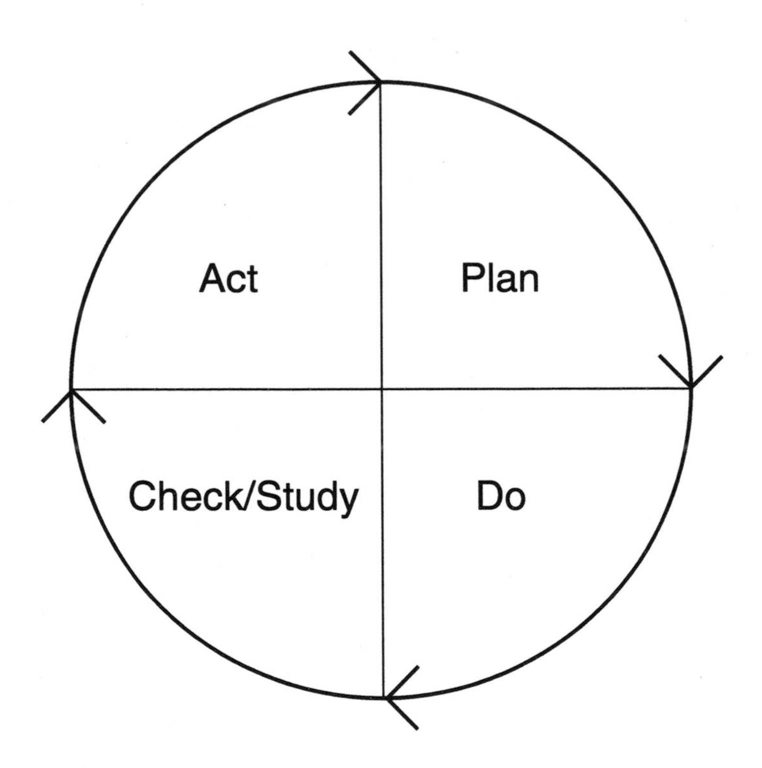

Figure 14-1 Basic Plan-Do-Study-Act Cycle.

Any activity resulting in learning or improvement, whether individual or group, is a candidate for the PDSA approach. The first step, Plan, is the planning stage for correction or improvement. Document the Plan so that a reference exists for use during the Do and Study stages. The Do stage is the execution of the plan. During this stage data is collected relating to critical aspects of the outputs of the process. The Study stage uses the data from the Do stage and compares this data with the plan. The Act stage drives the planning process for Plan or Do improvement. The Act stage drives the standardization process. Standardization makes certain that all who have a stake in the correction or improvement are included. educated and take ownership.

Shewhart and Deming both warned that a process must be in control before improvement is possible. In other words, if the results of a process are unpredictable, this unpredictability may be working against appropriate improvement or may mask undesired change. Early PDSA efforts must be to bring the process output to a high degree of predictability, even if the output is not exactly what the next process or the customer requires. Therefore, the early PDSA cycles are problem-solving attempts rather than improvement actions. Once the process is in control, use PDSA to improve the process to meet or exceed the requirements of the process.

One minor problem with the use of PDSA is that the cycle does not really start with planning, as "PDSA" would imply. Something must produce the need for planning. The need for planning occurs in a check or study activity. Something has not gone as anticipated. This drives an action step. Action results in planning. Develop a plan and execute it. Study the results to detect the need for further action, whether to make additional corrections or more improvement. A graphic presentation of the actual PDSA, along with the actual steps that take place in applying PDSA, can be found in Figure 14-2.

An excellent example of the use of the PDCA cycle is the improvement of team meetings. A meeting agenda calls for a one-hour meeting (Plan). The meeting takes place and lasts one hour and twenty minutes (Do). At the end of the meeting the team leader suggests that there is a discrepancy between the planned length and the actual length of the meeting (Check). The team discusses the number of agenda items, time required for each, tangential conversations and other data relating to the meeting (Study). The team decides that improvement in how they manage the agenda is necessary and that they must eliminate side conversations (Act). This guides the team to making a commitment to keep discussion on track during the next meeting, allot more time per agenda item and watch time more closely (Plan). At the conclusion of the next meeting (Do), the team discusses how their plan worked (Study) to decide if further meeting improvement planning is necessary (Act).

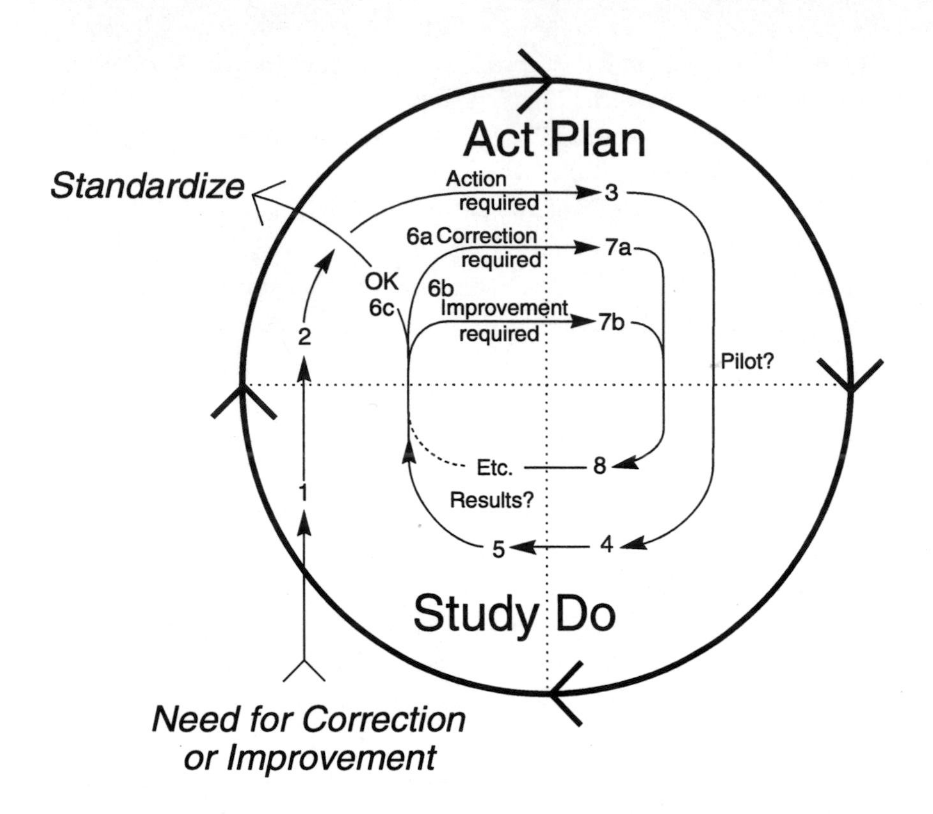

Figure 14-2 The PDSA cycle is actually initiated with the recognition of the need for correction or improvement.

PDSA is an approach used for both problem-solving and continuous improvement activities, following a step-by-step process:

1. A discrepancy between the plan and the results observed, drives a STUDY of what is causing the undesired effect. Data is collected, analyzed and interpreted.
2. The team determines that some areas of improvement or change are necessary. They decide to ACT on the data to cause improvement.
3. The team develops a PLAN to address root causes and implement improvement, thus closing the gap between what they want and what actually occurs.
4. The team implements the plan. They DO it. Understand that the best plan, poorly implemented, will probably not produce desired results. Also, a poor plan, well implemented, will not close the gap between plnned and actual.
5. STUDY the effects of the plan implementation (pilot, test, full-scale application).
6. ACT on the study results:
   a. If implementation did not close the gap between desired and actual, action calls for major revision of the plan, implementation, or both. The data from the pilot is additional information for study and should not be considered as information indicating failure. Give additional attention to identification of root causes. Caution, an excellent plan, poorly implemented will provide data suggesting a defective plan.
   b. If the results suggest further refinement of the plan, then the action would be to improve the plan, based on the results of the pilot or test.
   c. If the results suggest an effective plan, well implemented, then ACTION would be to standardize the changes or improvements through documentation, communication and education.
7. Second stage planning:
   a. PLAN to correct a disappointing pilotby including major modifications in the planning or implementation process. A completely new plan may be necessary in isolated cases
   b. PLAN to improve the original plan for a second pilot should include improvements rather than major modifications or a completely new plan.
8. Conduct pilot (another DO).
9. STUDY the pilot results . . . and so on, until the activity, process, system or product at least meets, preferably exceeds customer requirements.

Chapter 15

Improvement Tactics

Improvement

Much of the improvement and innovation we enjoy today is the result of learning and the implementation of this learning. Chapter Fourteen considers the basis for most learning, the P-D-S-A cycle. Improvement also comes from the application of what is already known, but is not used, applied properly or used to its greatest extent. Applying what is known to new situations is another approach to improvement.

This section provides an alternate approach to process improvement through an application of the P-D-S-A learning cycle: Plan-Run-Study-Act, or Precontrol. This approach forces process improvement through feedback. Precontrol, and P-R-S-A, couples the power of statistical approaches and the continuous learning cycle to process improvement.

The remainder of the chapter offers suggestions for improvement tactics: Mistake Proofing, Actions on Systems and Local Causes of Variation. Finally, this chapter provides an Implementation Checklist to make certain a high quality plan for improvement is implemented correctly.

Precontrol/Plan-Run-Study-Act

Precontrol is an application of statistical process control (SPC) to force a process in control to meet or exceed customer requirements[1]. Precontrol is statistically sound and allows little chance of permitting a process to continue while it is out of control.

Precontrol is an extremely simple method for process operators to use as they gauge how well their processes are meeting the requirements of subsequent processes or customers. Precontrol begins by dividing the tolerance band (the area between the upper and lower specifications for the measurement) into three zones: a target zone (green) bounded by two cautionary (yellow) zones. The first phase of Precontrol is to bring the process in control and meeting the specifications imposed on the process. Then the operator periodically takes two samples to learn how the process is doing. If both samples fall into the caution zones or either sample falls out of the tolerance band, the operator stops the process and implements a correction. If one falls into a caution zone, the operator takes another sample to make certain that the first sample in the caution zone was simply a random event.

The theory supporting Precontrol is based on the normal distribution. One and one-half sigma on each side of the target value (or the midpoint between the upper and lower specification) will include about 86% of the expected occurrences when the process is in control. In other words, 14% of the time a sample will fall in one of the two yellow zones. If two sequential samples are taken from an in-control process, then the operator could expect both to fall in the yellow zones only about 2% of the time (.14 times .14 equals .0196, or 1.96%). Thus, if the operator gets two sequential samples in the yellow and stops the process to learn what the problem is, the operator will be wrong in stopping the process only about 2% of the time. The other 98% of the time there is indeed something wrong. The risk also exists that the process has gone out of control, but the operator, by chance alone, has chosen the two samples which happen to fall in the green area. The operator can expect this to occur about 1.5% of the time. It is unlikely that two samples falling in the green zone would be chosen at the next sample time. Exposure to this risk can be reduced by decreasing the time between samples.

[1] Precontrol was developed in 1953 by Warren R. Purcell, Franklin E Satterhwaite, C. W. (Bill) Carter and Dorian Shainin from the management consulting firm of Rath and Strong, Inc. Under contract from the Jones & Lamson Machine Co., they were charged with a more direct and simplified approach to statistical process control, as developed by Walter A. Shewhart. The first company to use their new approach was IBM at Endicott, New York in 1954. Unfortunately, the economic success of the fifties and sixties relegated the use of precontrol and SPC to little used tools. SPC became fashionable again in the early eighties when U.S. Organizations observed extensive use of SPC by Japanese manufacturers. Precontrol was not recognized as a powerful tool because of its simplicity. Precontrol is being utilized more now as companies become disenchanted with the complexity of SPC and the need to continually measure output, frequently.

Precontrol also allows quick calculation of Cpk. The Cpk of a process suggests to what extent the process is within the specifications imposed on the process. A Cpk equal to one suggests that the process is just meeting the requirements of the process. Statistical process control and other approaches require many samples to be certain of the Cpk of the process. The operator need not take many samples to determine the Cpk of the process, once in control. Setting the green zone range within the +1.5 sigma to -1.5 sigma range produces a minimum Cpk of 1.33. There is a 99% chance that if the process slips below Cpk of 0.8 or less, the next sample period will indicate shutdown. If the process runs for some time without a sample measurement in the yellow zone, then the process is running at a Cpk of 2.0 or more.

The Precontrol Process

Precontrol is essentially the opposite of statistical process control. Precontrol begins by imposing the specifications on the process rather than determining what the process can produce. After implementing a process improvement, the operator takes a minimum of five samples at short intervals while "tuning" or improving the process. This activity continues until the process is producing output within the imposed specifications as indicated by the Precontrol decision rules. The activity of improving the process to produce conforming output might require a variety of problem-solving or process-improvement approaches. The five-sample measurements provide valuable feedback during the problem-solving or process improvement period. When the process is producing conforming output and, therefore, is in control, the operator begins taking two-item samples at longer intervals to confirm that the process remains in control and to specifications. The operator determines the period between each double sample by making note of how often stoppages (two yellow zone samples or a red zone sample) occur. Divide the time between stoppages by six. This is how often the operator should take samples. If the operator wants a greater amount of protection, divide by 10 or more. The operator may desire greater protection to reduce the chance that the process is out of control but not detected, or to shut down the process as soon as possible when it goes out of control. Dividing the stoppage interval by 24 is the practical upper limit.

Though a process producing goods is much easier to improve using Precontrol, processes producing a service are candidates for Precontrol. The requirement ia that the specification applied to the process must be discrete or continuous (see Chapter One). Time is an appropriate specification for a services process as is volume or frequency. Chapter Two includes several examples of service process measures that could be used with the Precontrol approach.

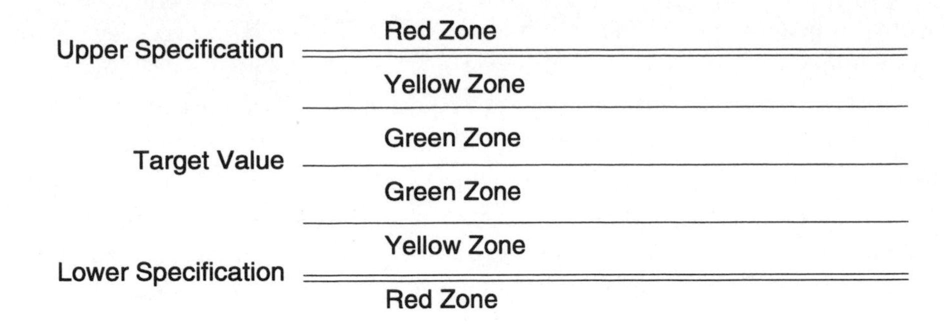

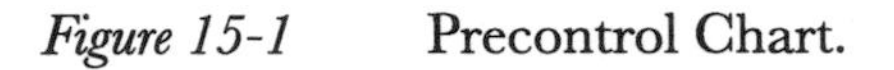
Figure 15-1 Precontrol Chart.

Decision Rules:

I. Process setup and qualification requires 5-item samples. Process is in control when five samples in a row produce measurements in the green zones. If one yellow occurs, restart the count. It two yellows in a row occur, make an appropriate process adjustment and restart the count.

II. Process control: Once the process is in control and adjusted to output specifications, take two item samples.
- If both fall in the green zone, continue to run the process. (Expected about 73% of the time.) Five sample pairs in a row, all falling in the green zone, decrease the chances of missing a yellow or red zone sample almost to zero.
- If a sample falls in a yellow zone, continue to run the process. (Expect this to occur about 25% of the time.) If the three pairs in a row each produce one measurement in the yellow zone, make a process adjustment, (This will occur naturally about 1% of the time.)
- If two yellows occur in the same yellow zone, make adjustment. (This can occur only about 1% of the time when the process is in control.)
- If two yellows occur, one in each yellow zone, stop the process and call for assistance, as this calls for more sophisticated correction. (This will occur only about 1% of the time in a process in control.)
- If one or both measurements fall in the red zone, stop the process! Defects are occurring. Check output produced since last samples. (The probability of this happening while the process is in control and meeting specification, is extremely small.)

Note: Remember, the risk of stopping a process when it should not be stopped is about 2%. This is better than not stopping a critical process that is producing non-conforming output. The risk of not stopping a process that has gone out of control will not exceed 1% (for a six-pair run) and is typically near 0%.

Plan-Run-Study-Act

Plan-Run-Study-Act (P-R-S-A) is the application of P-D-S-A to process improvement utilizing Precontrol. Precontrol provides immediate feedback of the most recent Run of the process for the Study phase of the cycle. The Precontrol decisions rules provide statistically valid directions for taking action. The process operator can make changes anticipating output that will subsequently meet or exceed customer requirements. Continued sampling according the Precontrol rules will provide continued information on the effectiveness of the change, again, relating directly to customer requirements.

Once the process meets the specifications set for the process, Precontrol and P-R-S-A can be used to further improve the process to exceed customer expectations. Or, another, less critical specification can be chosen for improvement through Precontrol/P-R-S-A.

Mistake-Proofing

Many processes have been designed or allowed to evolve in a way that permits mistakes to occur. If the process is critical, or could affect subsequent critical processes or the customer, then Mistake-Proofing should be a factor in the process improvement activity.

An excellent example of Mistake-Proofing is the pilot's check list. The checklist moves the pilot through key measures, control positions and functional checks. Each item is checked off before moving to the next. This improves the safe operation of an aircraft, significantly. Another Mistake-Proofing activity is the "to do" list kept by many people. Forgetting a critical activity is all too easy.

Though check lists and "to do" lists are excellent examples of Mistake-Proofing, many other approaches are available. Placing the numbers in a different arrangement mistake-proofed a telephone keypad. The reason for the difference is that those people quite skilled at entering numbers quickly on the calculator keypad could key telephone numbers faster than telephone equipment could respond. Many wrong numbers could be keyed in and the reversed keypad layout eliminated this problem.

Mistake-Proofing can be as simple as requiring a counting step in a process. For example, sponges are counted before surgery and after to confirm all have been removed from the patient. Another simple Mistake-Proofing scheme is the link between the automatic transmission shift and the break pedal. The transmission cannot be shifted from "park" to "reverse" or "drive" unless the brake pedal is fully depressed. This virtually eliminates the possibility of sudden acceleration when shifting into gear. Check Lists can be computerized. An example of a type of computerized Check List is an order entry screen on a computer. The order entry operator must follow the prescribed order and fill blanks in a specified manner.

Procedure for Mistake-Proofing

Begin by identifying critical processes and activities where mistakes are occurring or could occur. If using a team approach, the team should set priorities for the processes to be Mistake-Proofed through Multivoting. If individual, consider the outcome of individual mistakes. First, Mistake-Proof those processes that could allow mistakes resulting in high correction costs, potential damage to or defects in later processes or pose a threat to human life or health. Then, follow these steps to mistake-proof the process or activity:

1. What are the critical elements of the process or activity? What does the customer of the process want and sometimes does not get? What, specifically, sometimes goes wrong? All answers are candidates for Mistake-Proofing.
2. What approach might eliminate, or at least drastically reduce mistakes? The possibilities include:
   - Check lists
   - Computer assistance Forced entry fields, numerical only fields, spelling and grammar checkers, warnings of out of expected limits for an entry, etc.
   - Technology Often, the application of new technology and materials will overcome the tenancy for mistakes. Universal Product Codes (UPC), have virtually eliminated keying in the wrong price. Now the Mistake-Proofing activity must move to the accuracy of the code itself. Ground fault detectors now protect people from mistakes associated with using electrical appliances in a way that could cause electrical shock.
   - Assigned Responsibility When responsibility for a critical process or activity is not assigned to someone or is assigned to a group, things begin to "fall through the cracks." Sometimes it makes sense to assign responsibility to a single person. An example is bulletin boards. If no one person has responsibility for maintenance, the material on the board remains on forever, critical items are missing and a push-pin is not available when needed. In a bank, each teller has primary responsibility for their cash.
   - Simplification Upon simplifying a system, process or activity, far fewer errors occur. Forms that require entering the same information in several different places on several different sheets invite mistakes. Twenty digit part numbers are more mistake-prone than ten-digit numbers.
   - Feedback Transposition errors are all to common. Use feedback mechanisms to reduce transposition errors. When a caller gives a telephone number, street address, item order number or an account number, Mistake-Proof the process by repeating the number to the caller for verification. Confirm spelling by using the same process.
   - Visual Checks A picture is worth a thousand words when Mistake-Proofing. Diagrams, drawings, pictures and outlines are a few of the visual error reduction approaches. This is why pictures are included with a "some assembly required" purchase. Outlining the tools on a tool board virtually eliminates the possibility of not knowing a tool is missing. Better yet, paint the area hidden by the tool with a contrasting color such as orange. The absence of the tool will be annoying. The user is much more likely to replace the tool.

- Indicators and Alarms A variety of "indication" schemes can be used to "prove" or document that a critical activity has taken place. Automobiles have temperature gages to tell the driver that the engine is too hot. Often, these are ignored. Thus, engineers include "idiot" lights to get the driver's attention. Lights and alarms alert the driver and passengers to an unfastened seat belt. Computers have a variety of indicators and alarms, that help prevent mistakes that would destroy critical files. "Are you certain you wish to delete this file?" have saved many irreplaceable files.
- Logical Sequences If groups of sequential activities are not in a logical sequence, mistakes will increase. Make certain sequences of activities and processes are logical and make sense. Good cooks know that a logical sequence in preparing meals prevents mistakes. This is how they get all the food finished simultaneously. Effective memos require writing in a logical sequence. Get the information, outline how the memo will be written, write the memo, ask someone else proofread the memo, and then send the memo. This makes more sense than having someone proof the outline for spelling and grammar, then sending a critical memo out without proofing.
- Consistent Sequence Once a quality sequence has been established, stick to it. This reduces the chance that a critical step or process will be over looked. Developing the habit of always doing things the same way may seem boring and counter to quality improvement, but problems often follow change that is not for a good reason. (Again, good cooks do not constantly change the sequence of recipe steps). When the right sequence is found, the sequence becomes standardized. Check lists assist in following a standard sequence.
- Planning for interruptions Interruptions add opportunities for mistakes or skipped activities. All critical processes and activities should include a procedure for noting the point where an interruption has taken place. The note provides a starting point after the interruption. Better yet, processes and activities should be completed to a standard point before being interrupted. Critical processes should have contingency plans for interruptions. Bookmarks are simple examples of planning for interruptions. Some readers follow the rule-of-thumb that a chapter is always completed before putting the book down. Many work processes should be completed to a logical stopping point before taking a break, going to lunch or leaving for the day. Post-it® notes help keep one's place when someone interrupts a work process, whether in person or by telephone. Use a bright-colored note for emphasis.

- Write It Down The human mind is remarkable, but can be fallible. Forgetting critical facts is quite easy for most people. This is the reason for a daily planner, the "while you were away" notes, and a shopping list. Voice-mail and E-mail can provide documentation of facts. One interesting use of voice-mail is for a person to call himself or herself at home or the office, leaving a message concerning what should be done upon arriving at that location. Pocket voice recorders are great Mistake-Proofers.
- Label Similar items with critical differences could be interchanged. Label these items in a way that distinguishes between them. Color coding, shape, textures, symbols, numbers, words or pictures work nicely. Where would we be without the odor added to natural gas to "label" it? Many pills look the same, once outside their containers. Pharmaceutical companies label pills in a variety of ways to help prevent mistakes. The "f" and "j" keys on most key boards have a raised dimple to help the fingers in finding the proper positions on the key board.
- Locators Often the assembly or use of items is possible in two or more orientations or positions. Locating pins, dimples or keys can be added to prevent all orientations or uses except the proper one. Most computer cables have locating pins so that insertion is possible in the proper orientation, only. Electrical plugs have one connection larger than the other to prevent the plug from being inserted the wrong way.
- As a last resort: Make Major Changes in the process or activity. Rethink or reinvent the process or activity so that the expected mistakes cannot occur. The telephone keypad was rethought. Many mistakes (resulting in accidents) were made at highway intersections. Stop signs, stop lights and reduced speed limits were not the answers. The interchange was the product of rethinking the way two highway cross without reducing vehicle speed to any extent. The straight pin was reinvented and became the safety pin.

3. Test or pilot the improvement to confirm that mistakes will not take place.
4. Standardize (record, educate and communicate) the improvement.

Some traditional activities, thought to be Mistake-Proofing, are not. Proofreading and other inspection schemes add cost and reduce the throughput, but do not eliminate or reduce all mistakes. In fact, errors increase since mistakes are not critical. After all, the inspector will discover the mistake. These approaches do catch some, possibly most, of the mistakes after they occur. Then the mistake must be corrected, which could introduce problems with another part of the process. Watching process operators very closely will not eliminate mistakes either. This only frustrates the operator and increases the opportunity for errors. Neither should a job be "idiot proofed." "Idiot proofing" involves making a job so simple, requiring so little thinking, that even an "idiot" could do it. These are not jobs for humans and are not an appropriate way to Mistake Proof.

Process Simplification Approaches

Simplifying products, systems, processes and activities is one of the best ways to reduce problems, and improve quality and effectiveness. Process Simplification can take several forms:

1. Eliminate unnecessary steps, steps that do not add value. Inspection, moving to and from storage, storage itself, sorting, and rework are examples of non value-added steps. Unnecessary approval steps is another form of complexity.
2. Flow Chart the process. Now, rearrange the work activities in a more logical, simpler order. The goal is to remove unnecessary steps, branches and loops in the process. One approach is to use Post-it® notes of different color. Pink (red) indicates that which is required, yellow indicates that which should be included and blue indicates that which is unimportant or is not required. The simplification activity focuses on the blue notes first.
3. Construct a Reverse Fishbone to learn what could cause complexity in the process. Those parts of the product, system, process or activity that cause complexity and do not contribute to value or quality should be suspect.

Differentiating Actions on Systems and Local Actions

Variation and errors come from two general, and very different, sources. Some variation and occasional errors (non-conforming output) are part of every process. This is the natural variation in the process. The variation is built in. It cannot be attributed to any single factor and is reduced by making the process more robust, taking out that which contributes to the variation. Natural variation is what produces the normal distribution or the bell-shaped distribution curve.

Variation also occurs in processes due to special or assignable causes. Identifying these causes and attributing the variation directly to these causes is possible. Usually linking these causes of variation to individual people or individual aspects of the process is possible. This is not to infer that people purposely cause problems, only that they were an intimate part of the process problem. Special causes distort the normal distribution.

An example will help. Dimes are minted to represent one-tenth of a dollar and to function in coin changers. Most coin changers (or at least a high percentage) will accept any U.S. dime deposited into the changer. Yet dimes vary. The variation includes weight, thickness, diameter, and appearance; all due to variations in the metal alloying and minting process, and wear. As an example, sampling and weighing the dimes minted in a given year will produce a normal distribution of weights. Most of the dimes will fall near a mean weight. Some will weigh a little more than the mean and a few will be even heavier. Some will be a little lighter than the mean and a few will be even lighter. The common causes of variation in the weight of dimes do not generally cause problems with people identifying dimes and using them in coin changers. The U.S. mint does not need to engage in improving the system which makes dimes. The system is robust enough, meeting the needs of the customers using dimes.

Nevertheless, the process that makes dimes can fail. When this occurs, the mint should attribute the failure to a special cause. For example, the equipment that forges the dime could experience a component failure and produce double-struck dimes. The component failure is a special cause of variation, attributed to the failed component. A "people" failure could be due to a replacement operator without proper training.
If coin changers and counters could not count dimes fast enough or accurately enough, then the system in which dimes are minted would require change. This problem would be classified as a common cause among dimes. Common causes of variation must be addressed by and within the system the process operates. Reduction of common causes of variation should be an ongoing activity in all organizations since it has become common knowledge that most variation is due to common causes of variation.

Even upon identification of a cause of variation as "special," the root cause could be common. Many quality experts have suggested that 80 to 95 percent of the variation in process outputs is attributable to common causes. This suggests that most variation reduction efforts should be directed toward common causes. In the double-struck dime example, lack of preventive maintenance or insisting that equipment continue operating with known component problems could be likely root causes of the component failure. Both are system problems. System problems require management actions. Treating a common cause as attributable cause demoralizes the people involved in the process. Disciplining the operator when a double-struck dime is produced is a poorly aimed corrective action, and will demoralize the operator. Disciplining the maintenance person in a "Don't replace it until it breaks" environment demoralizes other employees.

It is possible to track the special causes of variation back to individual (special) situations. These situations do not commonly occur. Someone must deal with each special cause on an individual basis, as close to the process as possible. This is a critical concept. Dealing with a special cause as a common cause will increase variation. Correction then becomes expensive.

An example might help in understanding the critical difference between dealing with common and special causes. If one employee violates the policy that employees are not to make personal photocopies, a special cause exists. Special action would be to put the offender on notice. A common cause action would be to install a video camera and video recorder. Or, require every employee to check photocopies with a copy room attendant. If most employees are making personal copies, counter to policy, then a common solution would be in order. Communicating the policy in strong terms might be one possible solution. The "copy room attendant" might be a solution if the problem is extensive.

Handling special causes with system solutions are too common and usually create more problems than improvements. Some supervisors carefully watch all direct reports just in case one employee decides to stray from the directed path. The supervisor should obviously be dealing with the offenders. However, some situations exist where the only alternative is to install system changes to deal with specific problems. Congress enacts highly restrictive laws specifically to deal with a few offenders. Airports install expensive security measures to deal with a few hijackers and terrorists.

Organizations often treat common causes as special causes or permit common causes to produce special actions. If an organization typically buys from the lowest bidder and the lowest bidder delivers defective material, the person who makes the product must cope with the problem. Often this person "pays the price" of this common problem through additional effort, increased stress and possible discipline actions.

Implementation Checklist

An Implementation Checklist should be completed before beginning a pilot test or full scale implementation of any change involving a risk of failure. The improvement team should go through each item to confirm that the team is prepared for the implementation phase of the project. Every success factor of the implementation phase must be considered. Often individuals and teams do not fully consider what could go wrong and make contingency plans.

Appendices

Appendix A

The Language of Quality Management

Activity: An observable portion of a process. Processes are made up of activities.

Consumer: The ultimate customer of a good or service, whether inside or outside the organization.

Critical Process: A process in a collection of processes or system that affects the suitability of the output considerably more than other processes. Critical processes must produce conforming output.

Crossfunctional: Improvement and problem-solving activites and team that include two or more functions within the organization.

Customer: The person, group or organization receiving a good or service, whether inside or outside the organization. "Customer" may be the next process, others in the organization, and includes immediate customers as well as the ultimate consumer.

Improvement: The result of a process becoming more effective. The output of a process better meets or exceeds the requirements of the customer of the process or maintains the level at which the needs are met but utilizing less resources, or both.

Key process: A key process will affect the quality of subsequent processes much more than other processes.

Manager: Traditionally, a person who plans, organizes, implements, controls, and has control over a functional area. Often a functional expert or professional. The manager's new role is to concentrate more on clearing barriers for those in his or her functional area, develop people, garner the resources necessary, sponsor teams in the functional area, pass communications downward and upward with minimal distortion.

Organization: A formal collection of people with the ultimate objective of providing a good or service to external customers. Organizations include for-profit goods and services companies, non-profit companies, governmental groups, public service groups, clubs and associations.

Output: That which a process or system produces. The output of one process moves to other internal processes (internal customers) or external customers (to be inputs to their processes). Includes goods, services and information. Often, a product thought of as agood, only, also includes a service. For example, the sale of an item could include training on the use of the item, support on installation and after-sale support. A product thought of as a service could include goods. Though insurance is basically a service, the goods portion is the actual contract.

Process: A process takes inputs and transforms these inputs into a good or service to meet the needs of the next process, other customers or ultimate consumer.

Process Resources: Anything utilized by the process to produce an output. Includes time, human energy, money, materials, information, space, etc.

Product: The goods or services provided by a person, group or organization. Products are the outputs of organizations and include information.

Quality: How closely a good or service matches the expectations of the receiver of that good or service. The sum of the features and characteristics of a good or service which contribute to its ability to satisfy a need.

Quality Assurance: The group responsible for preventing the delivery of non-conforming product through a variety of prevention tactics.

Quality Control: The group responsible for tracking, recording and reporting the quality of the organization's product. Also responsible for sorting out the non-conforming and approving the conforming product.

Quality Management: The new way of managing an organization or group within the organization characterized by satisfied employees, customers and stakeholders.

Requirements: What subsequent processes, internal or external customers, consumers must have of the process. Less critical needs, often referred to as "wants" do not become requirements until critical needs are meet, even exceeded.

Specifications: A set of expections set for a product, system or process. Specifications are set by the consumer, external customer, subsequent process, an expert, industry standards or by governmental regulations. Specifications are most often stated in association with statistical process control charts. Specifications are also used in advertising literature and literature accompanying the product to inform the customer of what the product is designed and produced to provide to the customer.

Supervisor: Traditionally, a person who has direct reports and accomplishes results through people. Supervisors, as with managers, are taking on new roles in organizations, distancing themselves from day-to-day decision-making and control activities.

System: A collection of processes responsible for a completed output.

Systems Thinking: Thinking of the organization as a collection of systems or one complex system functioning together to deliver customers or stakeholders the required goods or services. Systems thinking is contrasted to thinking about an organization in terms of its organizational chart or the organization as a collection of functions.

Team: A group of people working together toward a common goal, whether on an ongoing basis (work teams) or a project basis (cross-functional problem-solving teams). "Team" implies that the supervisor is a leader rather than the traditional "boss".

Value-added: That which increases the value of a good or service to subsequent processes, the ultimate customer or the consumer/user of the product.

Appendix B

Common Statistical Terms and Abbreviations

Term	Symbol	Definition
Average		Most often, average is meant to be "mean." Sometimes, however, "average" is used to signify median, mode or mean.
Average of the Averages (X bar bar) Grand Average	$\overline{\overline{X}}$	Add up the subgroup averages (sum of the $\overline{X}$) and divide by the averages of the number of divide by averages the number of subgroups (k).
Average of the Ranges (R bar)	$\overline{R}$	Add up all the ranges and divide by the number.
Constants for Control Charts:	A_2 d_2 D_3 & D_4	Factors for control limits. Factors for estimated standard deviation. Factors for control limits.
Estimated Control Limits for R	UCL_R LCL_R	D_4 times $\overline{R}$. D_3 times $\overline{R}$.
Estimated Control Limits for $\overline{X}$	$UCL_{\overline{X}}$ $LCL_{\overline{X}}$	Upper Control Limit: $\overline{\overline{X}}$ + A_2 divided by $\overline{R}$. Lower Control Limit: $\overline{\overline{X}}$ - A_2 divided by $\overline{R}$.
Estimated Sigmas	$\hat{s}$ $\hat{\sigma}$	Average of the ranges divided by d_2. The symbol over the sigma is a "hat." If the sample forms a normal distribution, is from a normally distributed population, and contains a large number of samples (perhaps more than several hundred), estimated sigma can be calculated by dividing the range of the data by six. This highly estimated sigma should not be used for critical decisions and is noted by two "hats."

Lower Specification Limit	LSL	Lower acceptable limit set by the customer or an "expert."
Mean (X bar)	$\overline{X}$	A measure of central tendency. Add up the scores (sum of the X's) and divide by the number of measurements made (n).
Median		The value such that half the values are larger and half are smaller. The median is the middle value. The use of "median" to note the central tendancy of the data will minimize the effect of outliers.
Mode		The most frequently occurring value or values. In a frequency distribution, if two peaks are observed, the distribution is bimodal.
Outlier		A sample or value significantly different from the others. Outliers are usually caused by special events, or a measurement or recording error. To better understand that being studied, remove the outlier from the data-set.
Population	N	Total number the "statistics" represents. The total number of cases (e.g., all customers) from which the sample was drawn.
Range	R	A measure of dispersion, and is the difference between the largest value and the smallest value
Sample	n	Number studied (from the population) to provide the data.
Sample measurement	X_i	The individual measurements (X_{first}, X_{2nd}, X_n, etc.).

Standard Deviation (Sigma)	σ	A measurement of variability.
	S	Standard deviation of the population.
	s	Sample standard deviation. Sum of the squared differences between each measurement and the average, which is then divided by the number minus one, the result of which a square root is taken.
	s_{-1}	This symbol is occasionally used to indicate that the sigma is for the sample population. The "-1" indicates that the sigma is for a data set with a degree of freedom one less than the number of values in the data set.
Subgroups	k	The number of subgroups (size n) used to compute control limits for control charts.
Upper Specification Limit	USL	Upper acceptable limit set by the customer or an "expert."

Appendix C

Improvement Tool Matrix

Need	Tool	Page
Converting Data to Visual Information	Histograms	76
	Pareto Charts	77
	Run Charts	79
	Scatter Diagrams	82
Deepening the Understanding of Information	Matrix Diagrams	106
	Stratification	103
Directing the Improvement Effort	Four Res	137
Feedback from the Process	Attribute Charts	45
	Decision Rules	51
	Run Charts	79
	XBar&R Charts	37
	X&MR Charts	43
Gathering Data for the Customers	Focus Groups	18
	Interviews	15
	Sun Diagrams	13
	Surveys	16
Gathering Data from the Process	Check Sheets	9
	Incident Charts	10
Generating New Information from Existing Knowledge, Improving Creative Thinking	Brainstorming	169
	Cause and Effect Diagram (Fishbone)	201
	Change of Entry Point	175
	Five Why's	198
	Indirect Analogy	173
	Random Stimulation	172
	Reverse Fishbone	171
	Reporter's Six	206
	Shifts in Attention Area	175
	Silent Brainstorming (Modified NGT)	168

Index